Apple®
Watch

2023 Edition

by Marc Saltzman

Apple® Watch For Dummies®, 2023 Edition

Published by: **John Wiley & Sons, Inc.,** 111 River Street, Hoboken, NJ 07030-5774, www.wiley.com

Copyright © 2023 by John Wiley & Sons, Inc., Hoboken, New Jersey

Media and software compilation copyright © 2023 by John Wiley & Sons, Inc. All rights reserved.

Published simultaneously in Canada

For general information on our other products and services, please contact our Customer Care Department within the U.S. at 877-762-2974, outside the U.S. at 317-572-3993, or fax 317-572-4002. For technical support, please visit https://hub.wiley.com/community/support/dummies.

Wiley publishes in a variety of print and electronic formats and by print-on-demand. Some material included with standard print versions of this book may not be included in e-books or in print-on-demand. If this book refers to media such as a CD or DVD that is not included in the version you purchased, you may download this material at http://booksupport.wiley.com. For more information about Wiley products, visit www.wiley.com.

Library of Congress Control Number: 2022948919

ISBN 978-1-119-91260-6 (pbk); ISBN 978-1-119-91262-0 (ebk); ISBN 978-1-119-91261-3 (ebk)

SKY10038205_111122

Contents at a Glance

Table of Contents

Introduction

’m excited to present you with *Apple Watch For Dummies*, 2023 Edition — your definitive guide to unlocking the power of your smartwatch.

In this book, you find out how to take full advantage of Apple Watch's many features, all in a language you can understand. You don't need a degree in electrical engineering to follow along with this book (*badum bum!*). Whether you're tech-shy or tech-savvy or perhaps somewhere in between, my goal is to teach you — in plain English — how to master your new gadget.

In other words, this book is meant to break down geek-speak into street-speak!

Apple Watch For Dummies, 2023 Edition, covers all the things you can do with your sleek wrist-mounted companion, ranging from productivity and connectivity features to information, personalization, and navigation; health and fitness applications; entertainment options; and much more. And, of course, this book contains many visual examples of what you can do and what your watch should look like in certain situations, regardless of the model (or "Series") of watch you own.

About This Book

Not to belabor the point, but I wrote this book with one focus in mind: to cover all you need to know about Apple Watch — without the technical jargon you don't need.

After all, technology can be confusing, especially when it's a teeny device you wear on your wrist. Therefore, consider this book your definitive guide to unlocking Apple Watch's capabilities. If you're more of an intermediate to advanced user, however, I also include a number of tips and tricks on how to get the most from your Apple Watch. Also, feel free to experiment, which can be part of the fun, but it may take some time to figure out all it can do.

You'll need to get accustomed to the limited screen size, interacting with the watch via your fingertips, voice (or even by making gestures with your arm), and what apps work best on your wrist (just as you'll discover some things are simply better on a phone). Because Apple Watch has a somewhat steeper learning curve than other technology, give yourself a while to master it.

Also keep in mind that Apple adds new features to Apple Watch over "time" (sorry for the dad joke) or when a new operating system update is available. But don't fret: This book covers not only the basics but also advanced capabilities. And hey, it's the sixth edition, after all, and I'm keeping on top of what's new so you don't have to. And after you learn — and apply — a good number of the tasks discussed in this book, you should no doubt be comfortable with whatever new things the watch can do in the future.

How to Use This Book

Although this book is meant to be a handy and informative resource, I hope you find the tone conversational. And as with other *For Dummies* books, you can read *Apple Watch For Dummies* in any order you like. I suggest you start with the first chapter or two in order to learn the various parts of the watch and its user interface, but after that, feel free to jump from chapter to chapter if one topic interests you more than another. Perhaps start by thumbing through the specific topics in the Table of Contents and then go to a particular chapter that piques your curiosity.

For example, you might wonder about the fitness capabilities of Apple Watch — how it counts your steps and the number of stairs climbed, calculates distance traveled, determines your calories burned and heart rate, and so on — so you can turn to Chapter 8 right away. Or maybe you're anxious to master text messages, emails, and calls on your new device? That would be Chapter 5 — on keeping in touch with those who matter. On the other hand (wrist?), Chapter 10 focuses on using Apple Watch for making mobile payments at retail stores by waving your wrist over a contactless terminal to complete a transaction. You get the idea. Each chapter can stand on its own.

And if you're "old school" and would like a more linear read, go ahead and flip through it from beginning to end. Just don't expect a plot twist near the end!

In some cases, I cross-reference subjects with topics from other chapters whenever relevant, but you can skip over them if you like, or you can pursue them. I also cover how to best use your voice instead of your fingertips — after all, I wrote the book *Siri For Dummies* (shameless plug alert!).

It goes without saying that you'll benefit most from this book if you have your Apple Watch with you, along with your nearby iPhone — which may be required for some of the changes you'd like to make to Apple Watch — and if you ensure the battery is full on both devices. Oh, and it doesn't matter which Apple Watch you own — such as the original version, the more attractively priced Apple Watch SE, or the latest and top-of-the-line Series 8 model and Apple Watch Ultra (an ultra-durable model for those who love the outdoors) — because this book is relevant to every model that has come out over the years. If it's a feature or setting tied to a specific Apple Watch, I call that out for you.

The various tips and tricks throughout the book — as well as some interesting tidbits — will help you get the most from your Apple Watch. For example, you can use your Apple Watch to control your smart home devices, such as compatible lights, a Wi-Fi thermostat, or smart door locks. Also, your watch knows the difference between a tap and a press. And did you know your watch can tap *you* with a slight vibration whenever a new message arrives for you to read? Or it reminds you to be conscious of your breathing throughout the day, and guides you along with relaxation exercises? You learn how to do that — and much more — throughout this book.

But you don't need to wade through these extra Apple Watch factoids if you prefer to stick to the basics. Most of this extra content is labeled as Technical Stuff (see the "Icons Used in This Book" section). Then again, you might be more interested in these "sides" than the main course. (I'm sometimes like that when I visit my favorite restaurant.)

Foolish Assumptions

When writing this book, I made only two major assumptions:

>> You own one of the Apple Watch products.

>> You want to know how to get the most from it.

The watch doesn't come with an instruction manual, so consider this edition of *Apple Watch For Dummies* the closest thing to one — and a whole lot more too, if I may say so myself.

Icons Used in This Book

The following icons are placed in the margins of the book's pages to point out information you may or may not want to read.

TIP

This icon offers suggestions to enhance your experience. Most are tied to the topic at hand, whereas others are more general in nature.

This icon reinforces the importance of information related to Apple Watch. You might consider bookmarking the page or jotting down the information elsewhere.

Apple Watch is a promising new wearable platform, but this icon alerts you to important considerations when using it, including health, safety, or security concerns.

This icon warns you about geeky descriptions or explanations you may want to pass on — but don't expect a lot of these throughout this easy-to-read guide.

Beyond This Book

We're almost ready to dive into this book so you can master your Apple Watch, but I want to make you aware of an excellent online resource: The *Apple Watch For Dummies* Cheat Sheet: This site offers handy tips on using your watch to meet your fitness goals and getting turn-by-turn directions. To find access to this cheat sheet, go to www.dummies.com and type "Apple Watch For Dummies cheat sheet" in the search box.

Where to Go from Here

If you've never used an Apple Watch — perhaps you bought this book in anticipation of purchasing one or receiving it as a gift — it might be best to power up the watch, turn it on, and follow the prompts to set it all up. Chapter 2 goes into this if you prefer to wait, or feel free to dive in with the watch before you fully crack the spine of this book. Your call based on your comfort level.

Regardless of which model you own — or plan on buying or receiving — you don't need to know anything to begin reading *Apple Watch For Dummies*, 2023 Edition. All you need is your willingness to learn this exciting new wearable gadget, which should help add convenience, speed, and style to your everyday tasks.

Ready to start? Turn the page . . .

1

Getting to Know Apple Watch

Set up your Apple Watch and discover its many features. Learn about its important parts, including the Digital Crown button and the side button.

Pair your Apple Watch with your iPhone and then learn about setting up a passcode and cellular connectivity (if supported), monitoring battery usage, and protecting your valuable investment.

Explore the many ways you can interact with your Apple Watch, including tapping, pressing, typing, and swiping, as well as conveniently accessing Siri with your voice to help you complete watch-related tasks (and much more).

IN THIS CHAPTER

» **Understanding the different Apple Watch models**

» **Discovering the many features of Apple Watch**

» **Navigating the Home screen**

» **Exploring different parts of Apple Watch**

» **Understanding wireless capabilities and sensors**

Chapter **1**

Watch This: Introducing Apple Watch

So are you excited or what?

You're a proud owner of the trendy Apple Watch. Or perhaps you purchased this book in anticipation of picking one up or receiving it as a gift. Either way, thank you for reading *Apple Watch For Dummies* 2023 Edition. This easy-to-read book has one goal in mind: to teach you everything you need to know about Apple Watch. With simple step-by-step instructions, clear images, and accessible tips and tricks, this book will help you gain the most from your new wearable gadget.

In this chapter, I walk you through the basics of Apple Watch to help you discover what this teeny wrist-mounted computer is capable of doing. You find out about the different parts of the watch — on the outside and inside — as well as the layout of the Home screen. From ways to interface with content on the watch to the hidden wireless technologies to integrated sensors that track your moves, you'll soon have a clear picture of the 21st-century magic you're wearing on your wrist.

It's a beefy chapter, so let's get to it.

Exploring the Apple Watch Collections

Apple Watch comes in a few sizes and configurations. For Series 1, Series 2, and Series 3 watches, you have a choice of a screen that's either 38 millimeters (about 1.5 inches) or 42 millimeters (roughly 1.65 inches).

For Apple Watch Series 6 and Apple Watch SE (a less expensive model introduced in 2020), sizes measure 40 mm (1.57 inches) or 44 mm (1.73 inches), but the watches have narrower bezels (borders) than their predecessors.

Apple Watch Series 7 (2021) and Apple Watch Series 8 (2022), has two sizes: 41 mm (1.61 inches) and 45 mm (1.77 inches). These watches introduced thinner bezels than all Apple Watches to date; therefore, the watch face is virtually all screen.

And finally, Apple introduced Apple Watch Ultra in 2022, featuring a super durable smartwatch with a larger 49mm case size, and other bells and whistles.

REMEMBER

You measure your screen from the top to the bottom, not diagonally — similar to how most screens in consumer electronics are measured (such as those on smart-phones and tablets).

Although you likely bought a watch before buying this book, note that a few versions of Apple Watch are available today (the latest Apple Watch Series 8, is shown in Figure 1-1), and you can purchase a few accessories to customize your watch. For a more extensive discussion of the Apple Watch collections, or for ways to persuade a friend or co-worker that they need an Apple Watch, visit `www.apple.com/watch`.

Excluding the various bands you can buy from Apple, the six Apple Watch options are

>> **Apple Watch Series 8:** The latest Apple Watch model (as of this writing) features the same design as the Apple Watch Series 7 (2021) but adds a new temperature sensor system (for insights into women's health), "crash detection" (through motion sensors and microphone), and a low-power mode that can squeeze up to 36 hours on one charge (with iPhone nearby).

>> **Apple Watch Series 7:** This 2021 model adds a bigger and tougher screen than its predecessors, faster wireless charging, all-new colors, an optional QWERTY keyboard for typing, and more. Choice of materials include aluminum, stainless steel, and titanium.

» **Apple Watch SE:** Much like the less expensive iPhone SE, Apple Watch SE — updated in the Fall of 2022 — is meant to give you premium features at a more affordable price. It includes a great-looking Retina display, new dual-core processor for faster performance, advanced sensors to track your movement, sleep, crash detection (2022 model), and more.

» **Apple Watch Ultra:** Introduced in the Fall of 2022, this larger (49mm) Apple Watch is designed for sporty and outdoorsy types, featuring a more ruggedized body (titanium case), precision dual-frequency GPS, one extra (and customizable) action button, crash detection, and up to 36 hours of battery life. This watch also has three specialized bands for athletes and adventurers.

» **Apple Watch Nike+:** Ideal for fitness types who like the Nike brand, this special edition Apple Watch (and special loop band) was designed to be your running partner. The watch synchronizes with the Nike Run Club app and Nike Training Club app. You can now add exclusive Nike watch faces to this edition of Apple Watch.

» **Apple Watch Hermès:** A partnership between Apple and Hermès, this fashion-centric watch includes bold, colorful (and extra-long wraparound) leather bands and an exclusive new watch face.

You also have a ton of choice when it comes to materials you want in an Apple Watch and what style of band to choose. With Apple Watch Series 6, you can go with aluminum, stainless steel, titanium, or ceramic. Apple Watch Series 7 and Apple Watch Series 8 introduced five aluminum case finishes, along with a range of new band colors and styles. See Figure 1-2. The super-durable Apple Watch Ultra is made with titanium and supports three specialized bands.

In the fall of 2019, Apple also announced Apple Watch Studio, a website that lets you choose a case and pair any band. Try it out for yourself at `www.apple.com/shop/studio`.

LOCATION, LOCATION, LOCATION

Despite there being nearly 200 countries on Earth, only 9 of them received Apple Watch when it debuted on April 24, 2015: Australia, Canada, China, France, Germany, Hong Kong, Japan, the United Kingdom, and the United States. Now dozens of countries sell and support Apple Watch — including models with cellular connectivity.

Figuring Out What Apple Watch Can Do

Some people may question why they *need* a smartwatch. Perhaps you traded your watch for a smartphone years ago and now wonder why you'd go back to the wrist. One word: convenience. Not having to carry anything is pretty darn handy, which you soon find out when using your Apple Watch. Simply glance at your wrist to glean information — wherever and whenever you need it — not to mention the fact that your watch can tap you with a slight tactile vibration to let you know about something, such as a calendar appointment or a loved one giving you a virtual "poke." Buying something at a vending machine or a retail store by simply waving your wrist over a sensor is also kind of awesome. Having an airline attendant scan a bar code on your watch's screen to let you board a plane? What a time saver.

Thus, you can keep your iPhone tucked away, preserving its battery for when you really need to access something with it. In fact, some Apple Watch models can make or receive calls and texts even without a smartphone nearby, which I get to in Chapter 5.

Perhaps because you wear it on your wrist and will likely glance at it multiple times throughout the day, Apple Watch will become an extension of yourself. When you strap it onto your wrist, you're not going to want to take it off. Now, that's personal.

As you discover in this book, Apple Watch has many, many features. Some of the main categories include time, communication, information, navigation, fitness, health and safety, entertainment, and finance (mobile payments). The following sections highlight Apple Watch's main features, but be aware that a few may require the GPS + Cellular model (and I indicate where).

Watch faces

Instead of a regular watch that simply shows one face, you can choose what you see on your Apple Watch. The watch has many styles to choose from right out of the box, as well as numerous downloadable apps that customize the look of the face. You can also change the color of the watch face to match your outfit. Chapter 4 walks you through it all.

Timers and alarms

Apple Watch also includes various stopwatches, timers, and alarms. Whether you use your fingertips or your voice, your Apple Watch can let you know when it's been 30 minutes so you can pull something from the oven. Or you can time your friend doing laps in a pool — from the comfort of your lounge chair. Apple Watch also lets you set an alarm to wake you up in the morning. You can use the Timer app as a game clock, for example, to tell you and your opponent when your time is up in a round of Scrabble. Check out Chapter 4 for all the details.

Caller ID or even calls

See who's calling by glancing at your wrist. Apple Watch displays the caller's name (Caller ID) or perhaps just a phone number (which often happens if that person isn't in your iPhone's Contacts). You can also use the Apple Watch microphone to record and send sound clips to friends. Some Apple Watch models — those advertised as GPS + Cellular — let you leave your iPhone at home and take or make calls right from your wrist when you're out! Heed the call, and go to Chapter 5 for details.

Walkie-Talkie

What's more fun and quicker than a phone call? The Walkie-Talkie feature built into Apple Watch. As the name suggests, Walkie-Talkie lets you press to talk to someone else who has an Apple Watch. Let go to listen for the reply. 10-4, good buddy? I cover this feature in Chapter 5, which is about different ways to use Apple Watch to communicate.

Health and wellness

In case you weren't aware, Apple Watch has been morphing into a powerful health device that can monitor what's happening inside your body. It sounds like science fiction, but the Apple Watch Series 6, Series 7, and Series 8 models include a heart-rate monitor (measured in beats per minute), electrocardiogram (ECG), and even a blood-oxygen monitor (to measure how well oxygen is being sent from the heart and lungs out to the rest of the body). Maybe not as impressive, Apple Watch can now detect when you're washing your hands — an important habit now more than ever because of the pandemic — and starts a 20-second timer. The new Mindfulness app features an enhanced Breathe experience, plus a new session type called Reflect. And watchOS9 adds a new Medication feature to its Health app.

Emergency SOS

In a nutshell, the Emergency SOS feature built into Apple Watch calls for help when you can't. Whether it senses a troubling anomaly through the heart-rate or ECG sensor or detects a fall, Apple Watch can dial emergency services, notify your contacts, send your current location, and even display your Medical ID badge for emergency personnel. I cover all this in Chapter 8, which focuses on fitness and health.

Similarly, Apple Watch Series 8, Apple Watch SE, and Apple Watch Ultra all add car "crash detection" (see "Crash detection," later in this chapter), and includes extra temperature sensors tied to women's health (see Chapter 8).

Text messages and instant messages

You can read and reply to messages with Apple Watch, as shown in Figure 1-3. Hold your wrist up to read the message and lower your arm to dismiss it. Chapter 5 walks you through all the messaging functions of Apple Watch, including some models that don't require your iPhone to be near you at all!

FIGURE 1-3:
Read and reply to messages on your Apple Watch.

Email

When an email comes in, you can read it on your wrist (scroll up and down the screen with your fingertip to see all the text), flag it as something to reply to later, mark it as read (or unread), or move it to the Trash. As with text messages and phone calls, you can transfer email from Apple Watch to your iPhone to pick up where you left off. I cover all this in Chapter 5.

Wrist-to-wrist communication

Along with providing the Walkie-Talkie feature, your smartwatch lets you communicate directly with someone else's wrist via a component called *Digital Touch*. Use your fingertip to draw something, such as the heart shown in Figure 1-4, and the person who receives it will see it animate — just as you drew it. Or why not send some virtual kisses to let someone know you're thinking about them? As described in Chapter 5, you can even send your heartbeat to someone by pressing two fingers on the screen.

Dock

Naturally, a wearable watch is a convenient way to stay on top of important information. Apple Watch has a cool feature called *Dock* that lets you quickly open your favorite apps or go from one app to another. To launch the Dock, press the side button and swipe up or down (or turn the Digital Crown button). See Figure 1-5. Chapters 3 and 6 discuss how to access Dock, customize what you see, and scroll through relevant information.

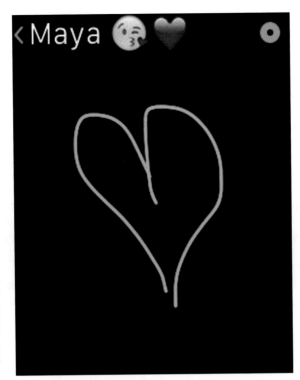

FIGURE 1-4:
Sketch
something on
your Apple
Watch and
send it off to
someone else's
Apple Watch.

FIGURE 1-5:
Dock allows
you to open
your favorite
apps quickly
or jump from
one app to
another.

Calendars

Apple Watch also has a Calendar app (with reminders) so you can stay on top of events occurring throughout your day (or coming in the near future). Also, when you receive a calendar invitation, you can immediately accept or decline it on your wrist and even email preset responses to the organizer. Put Chapter 6 on your calendar for more information.

Maps

Your wrist is an ideal place to glance at a map. Get turn-by-turn directions from your current location. You don't have to worry about having to stare at your wrist for visual cues (or fall down an open manhole in the process), because Apple Watch gives you a tap on the wrist to let you know when it's time to turn left or right. Navigate to Chapter 6 for more information. Beginning with Apple Watch Series 5, an integrated compass helps you navigate even further. (Unleash your inner Boy or Girl Scout Guide!) And with the latest operating-system update, watchOS 9, Maps includes cycling directions too.

Siri

Just as you can talk into your phone, Apple Watch has a microphone, which means that you have access to your personal assistant known as Siri. Flip to Chapter 7 to find out more about what Siri can do for you. As the author of *Siri For Dummies* (John Wiley & Sons, Inc.), I share some of my favorite Siri tips and tricks you can master with ease. Finally, in 2020, Apple added language translation to Siri's long list of capabilities. A redesigned Home app lets you control compatible smart home devices using your voice on your Apple Watch.

Fitness

One of the coolest applications for Apple Watch? Fitness. Chapter 8 looks at using the watch to measure your activity — steps, stairs, distance, time, calories burned, and heart-rate information — and to display it in a meaningful way on your watch and smartphone. I cover the Activity app, shown in Figure 1-6, and its three rings, which show you relevant information on your daily activity (or lack thereof!). On the other hand, the Workout app (shown in Figure 1-7) offers some workout routine options — including walking, jogging, running, and cycling — and shows real-time stats on your cardio session. Additionally, watchOS9 introduced new Workout options, so I take a look at those in depth, too.

FIGURE 1-6:
The Activity
app shows
three rings
that summa-
rize your daily
progress —
so far.

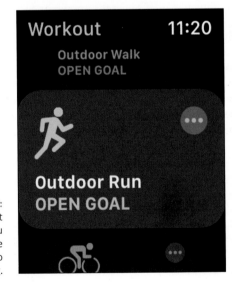

FIGURE 1-7:
The Workout
app offers you
some exercise
routines to
choose among.

Music connectivity and more

Chapter 9 shows you how to use Apple Watch like a wireless remote. Control your music on your phone — from the convenience of your wrist — as well as listen to synced playlists on your watch *without* needing your iPhone (but with Bluetooth headphones). Along with talking about music streaming and downloads, I highlight how to manage podcasts, audiobooks, radio plays, and other audio. Chapter 9 also covers how to control Apple TV on your Apple Watch. (Yes, you can pause and play episodes of *Ted Lasso*. How awesome is that?)

Apple Pay

Swiping your wrist at retail stores or at a vending machine is super-cool, and Chapter 10 covers all the ways you can use your Apple Watch in this regard. Your watch lets you buy products and services via Apple Pay, and you don't even need to have your iPhone with you.

Other apps

Apple Watch is quite a versatile gadget, which means that other apps can help enhance its convenience. Chapter 11 looks at optional third-party apps you can download to further personalize the most personal gadget in the world.

Other functions

Chapter 12 takes a closer look at some of the extra fun things you can do with Apple Watch. I cover using your wrist to snap a photo remotely on your iPhone, as well as look at photos on your wrist, including zooming in (because maybe you're bored in line at the supermarket and want to see some smiling faces or furry pets). I also discuss using Apple Watch as a gaming platform and tell you what's available for gaming.

Bonus tips

Chapter 13 reveals the top ten things you should try with Apple Watch and, of course, how to pull them off with grace. I share the absolute coolest things this smartwatch can do and how to best demonstrate them to your friends — to the point they'll be boiling with envy.

TECHNICAL STUFF

For Apple Watch Series 7, Apple Watch Series 8, and Apple Watch SE, Apple says you could squeeze up to 18 hours of battery life out of Apple Watch (see www.apple.com/watch/battery), but be aware that this result varies greatly, depending on how often you use the watch, the settings you choose (see Chapter 2), what apps you use, the outside temperature, and other factors. Apple says 18 hours equates to "all day" performance, which could include 90 time checks, 90 notifications, 45 minutes of app use, and a 60-minute workout with music playback from Apple Watch via Bluetooth. New in Apple Watch Series 8 is a Low Power mode option that disables some non-critical features and can double the battery life of Apple Watch to 36 hours (with iPhone nearby). Apple Watch Ultra also has a battery that lasts up to 36 hours, says Apple, and up to 60 hours in Low Power mode.

Determining What You Need for Your Apple Watch

The original Apple Watch didn't do too much on its own. Rather, it was more a companion device to an iPhone. Oh, sure, it could do a few things by itself — such as show you the time, count your steps, make payments, and play music — but a wirelessly tethered iPhone was required for the overwhelming majority of features.

With Apple Watch Series 8 and Apple Watch SE, you can go with one of two models:

>> **GPS:** This model is great for navigation.

>> **GPS + Cellular:** With this model, you can pay your mobile phone provider to unlock the eSIM (a virtual SIM card inside Apple Watch) so you can use the Apple Watch on the go like a phone. It can take calls, text messages, stream music, and more! In North America, this service costs $10 per month because it's added to your existing smartphone plan.

   Note: For Apple Watch Ultra, there is only the GPS + Cellular option.

This book is ideal for whichever model you have, so not to worry. As mentioned before, you do need an iPhone to set up Apple Watch, even if you have the version that doesn't require having one nearby to work. As you see in the next section, Apple also introduced Family Setup for those who own Apple Watch Series 4 (2018), or later. This feature lets family members who don't have their own iPhones, such as younger kids, still use Apple Watch to make phone calls, send messages, and share location information. (After you set up a watch for someone in the family, you can use your iPhone to manage some of that watch's capabilities.)

If you do own an older model (Series 1 or Series 2), you need at least an iPhone 5 to use Apple Watch. Those who own a Series 3, Series 4, Series 5, or Series 6 model need an iPhone 6 or newer. You also need to download and install the latest iOS operating system from Apple — whether you do it on your iPhone or via iTunes (on a PC or Mac) — and then connect the iPhone to your computer with a USB cable. After you download the latest operating system, an Apple Watch app — a white watch against a black background — appears on your iPhone's Home screen, as shown in Figure 1-8.

The Apple Watch app

FIGURE 1-8: Whether you own an Apple Watch or not, an Apple Watch app (shown at top right) appears on your iPhone's Home screen.

In fact, you can install the latest watchOS update over the air (OTA) without using a physical connection at all!

Getting to Know Apple Watch's Home Screen

As shown in Figure 1-9, the main Home screen of Apple Watch is populated by several small bubble-like icons. This screen is quite neat, actually, not to mention functional. Simply tap an icon with your fingertip to open an app or slide around the Home screen to see other icons pop up and grow larger. (You want the app to be centered on the screen for easy access.)

If you're an iPhone user, the icons should be familiar to you; therefore, you know what built-in and third-party apps launch when you tap a specific icon. Table 1-1 shows some of the built-in apps. See Chapter 3 for more on the native Apple Watch apps.

REMEMBER

You no longer need a nearby iPhone to install new Apple Watch apps as long as you're running the watchOS 7 operating system or later. (You can tell which version you're running by choosing Settings ➪ General ➪ About on the watch.) In Chapter 11, I cover both ways to install new Apple Watch apps, whether you want to use an iPhone (or Mac/PC) or the Apple Watch Store, or download directly to the device.

And, of course, third-party apps have their familiar icons, such as a big *P* for Pinterest, a swoosh for Nike, a green leaf for the Mint app, and so on.

TABLE 1-1

Built-In Apple Watch Apps

App	Icon
Activity	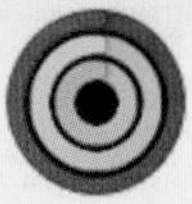
Alarms	
Calendar	
Find People	
Heart Rate	
Home	
Mail	
Maps	
Messages	
News	
Phone	

(continued)

TABLE 1-1 *(continued)*

App	Icon
Photos	
Podcasts	
Reminders	
Remote	
Camera Remote	
Settings	
Stopwatch	
Walkie-Talkie	
Wallet	
Weather	
Workout	

App	Icon
App Store	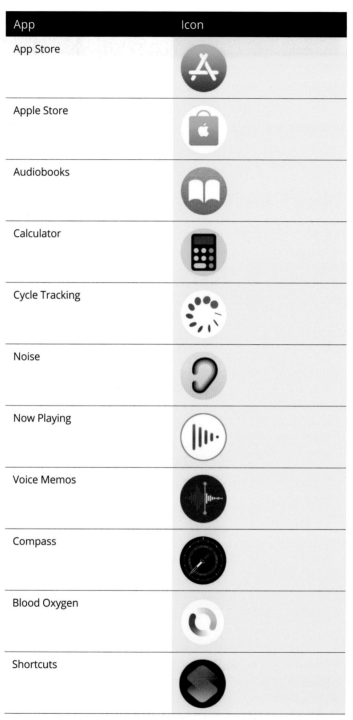
Apple Store	
Audiobooks	
Calculator	
Cycle Tracking	
Noise	
Now Playing	
Voice Memos	
Compass	
Blood Oxygen	
Shortcuts	

(continued)

TABLE 1-1 *(continued)*

App	Icon
Sleep	
Memoji	
Find Items	
Find Devices	
ECG	
Mindfulness	
Music	
Stocks	
Timers	
Tips	
World Clock	

App	Icon
Contacts	
Medications	

Even if you don't install any third-party apps, you'll be blown away at all the apps preinstalled on your Apple Watch, many of which you'll no doubt use. See Figure 1-10 for a sampling of what's already included.

FIGURE 1-10: Especially with the latest watchOS9 operating system, your Apple Watch can do so much.

Learning about Apple Watch's Parts

Okay, you're all geared up to test-drive all that Apple Watch can do, but if you're using it for the first time, you may not even know all the parts of the watch and what they do.

Fair enough. In this section, you discover the basics of the hardware itself. I start with a look at the various parts of the watch on the outside and what they do.

Watch face

Regardless of which size you opted to go with — 40, 41, 44, 45, or 49mm — the Apple Watch face is entirely digital, so you won't find any buttons of any kind. Use your fingertip to move the icon bubbles and tap an app to launch it. You can also tap, press, and swipe inside an app to perform a task.

WARNING

You don't need to press hard on these buttons or on the watch face. You want to minimize wear and tear on your new (and pricey!) gadget. Just a simple press on the buttons and watch face will do. And although Apple Watch Series 2 and newer are waterproof, try to avoid touching the screen and buttons with wet or damp hands. (Apple says, "We recommend not exposing Apple Watch to soaps, shampoos, conditioners, lotions, and perfumes as they can negatively affect water seals and acoustic membranes.") See Chapter 3 for more on these buttons and using your fingers with your Apple Watch.

Digital Crown button

Seasoned watch owners are familiar with the small rotary dial on a watch's right side (left-handed people may flip the watch around so it's on the left side), which is used to wind the watch (an old-school one, anyway) or set the time. Apple Watch has one too, called the *Digital Crown* shown in Figure 1-11. It's both a button and a dial, so it can be pressed, tapped, or turned forward or backward, with each change resulting in a different action. See Chapter 3 for more on what the Digital Crown button can do.

Side button

Along the side of the watch is a long button called the *side button* (how imaginative!), as shown in Figure 1-10 earlier in this chapter. From the Home screen and in any app, press this button to pull up your Dock (more on this later). Press and hold the side button to use SOS; double-click to use Apple Pay; or press and hold to turn your Apple Watch on or off.

Digital Crown

side button

FIGURE 1-11: If you wear Apple Watch on your left wrist, the side button is on the right side of the watch case. The Digital Crown button is the ridged dial.

The side button is flush with the side of the watch, except the larger Apple Watch Ultra, which protrudes a bit (which helps if you're wearing gloves outside).

Action button (Apple Watch Ultra only)

Among the many features of the Apple Watch Ultra is an extra button not available on the other Apple Watch devices. On the left side of the watch is an orange and pill-shaped "Action" button, which sits between the pinhole speaker and the siren (see below). This button is customizable, so it can perform a function you assign to it.

Until you change it in the Settings area of Apple Watch, pressing and holding the button will activate the 86-decibel siren that can be heard from up to 600 feet (180 meters) away, says Apple, to be uised in the event of an emergency. During workouts, the Action button will enable other fitness-related features.

Back sensors/charger

On the back of Apple Watch, as shown in Figure 1-12, are multiple sensors to monitor your heart rate, blood-oxygen level (on some models), and more. In fact, the addition of blood-oxygen monitoring beginning with Apple Watch Series 6 changed the back crystal underneath the watch; now it's made up of four LED clusters and four photodiodes. See Chapter 2 for more on the sensors and the charger.

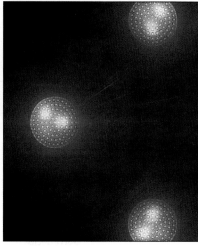

FIGURE 1-12: The sensors on the back of Apple Watch can, among other things, calculate your heart rate and measure your blood-oxygen percentage.

Watch band

Every wristwatch has a band to keep the screen snug on your wrist. You chose a specific band when you bought Apple Watch — a leather strap, a link bracelet, a classic buckle, a silicone band, and so on — but you can change bands later if you desire. Apple continuously introduces new bands, not to mention all the third-party ones available. It's all about selection and customization.

Using Apple Watch's Touchscreen

Just as you can interface with a smartphone, tablet, and laptop in different ways, based on the task at hand, Apple Watch gives you three ways to use the small screen on your wrist:

>> **Tap:** Tapping Apple Watch with one finger performs the same function that you'd expect on a smartphone: It selects whatever you're tapping, such as an icon to launch an app, a song to play a track, a link to a website, a photo to enlarge, or virtual buttons, such as on a calculator. On the Home screen, you tap and slide your finger around to move the icon bubbles. A tap is like a left-click on a computer.

>> **Press:** Apple Watch knows the difference between a quick tap and a longer press, usually when you need to open some additional menus. Think of a press as a kind of right-click. Tapping a song plays the track, for example, but pressing and holding it opens a set of options: Shuffle, Repeat, Source, and AirPlay. The technology that senses the difference between a tap and a press is called *Force Touch*.

>> **Swipe:** Many areas of Apple Watch — including Dock — and most of the apps you can access let you swipe left and right or up and down to navigate among screens. In Workout mode, for example, you can see time elapsed and heart-rate info, but swipe to the side to pull up music that you can pause and play. Swipe one more time, and you'll see more options, including the ability to lock your watch so you don't accidentally tap the screen during rigorous exercise, pause your counter, and so on.

Available only on the larger Apple Watch Series 7, Apple Watch Series 8, and Apple Watch Ultra is the option to pull up a small QWERTY keyboard (like a computer keyboard's layout) to type words. The keyboard's QuickPath feature also supports swiping from letter to letter to form words. Apple says that it uses machine learning to predict the word you're typing. See Figure 1-13.

FIGURE 1-13:
On Apple Watch Series 8, you can swipe between letters and let artificial intelligence predict your next words to speed your typing.

Some features are activated by two fingers pressed on the screen. In Chapter 5, you can find out how to record and send your heart rate or heartbeat to a loved one's Apple Watch.

TIP

Understanding Apple Watch's Wireless Functions and Internal Sensors

Oh, Apple Watch, you cleverly hide so much of your magic under your skin.

Apple uses an integrated computer described as a System in Package (SiP, for short) inside the Apple Watch. The SiP includes the main processor (the main engine that drives the watch's performance), along with memory, storage, support processors for wireless connectivity, sensors, and input/output tech. Yes, that's a lot of geek speak, which you don't need to know about to use the watch.

Beginning with Apple Watch Series 8, the company is using a new chip called the S8, which can provide wearers all-day battery life yet still power the larger, more advanced watch display.

Apple Watch indeed houses a good number of wireless radios beneath its surface, including Bluetooth, Wi-Fi, GPS, and NFC. To understand what they do, check out the following sections.

Bluetooth 5.0

Bluetooth makes a local wireless connection between two or more devices. Just as your wireless headset is paired with your smartphone so you can make hands-free calls, Apple Watch wirelessly communicates with a nearby iPhone to let you see texts on your watch, receive phone calls, control your music on your phone, and more. Bluetooth 5.0 works with devices up to 800 feet away (about 240 meters), which is significantly farther than in earlier versions. If you have an Apple Watch that supports cellular connectivity and pay for the service, you can perform many of these features — making calls, sending texts, and accessing online music — without having an iPhone nearby.

Wi-Fi

Even if you don't have a cellular model, Apple Watch features Wi-Fi, which gives it online connectivity even when no iPhone is in sight. As long as you're on a wireless network, such as your home's Internet connection or a coffee shop's hotspot, you can access such information as email, live sports scores, mapping information, and so on. A feature called *Continuity* — introduced in iOS 8 — allows you to receive messages and take calls on multiple iOS devices (such as answering a call on your iPad) as long as you're in range of your Wi-Fi network, and Apple Watch has this feature too. See Chapter 5 for details on how to take advantage of Bluetooth and Wi-Fi connectivity.

Cellular

As I mention previously in this chapter, newer Apple Watch models (Series 6, 7, 8, and SE) are available as GPS or GPS + Cellular. As you can guess, the GPS + Cellular watches cost a little more, but they let you make a call, send a text, and stream Apple Music from your wrist — all without your iPhone. You'll need to pay your mobile phone provider an extra amount per month (usually about $10) to activate the eSIM inside Apple Watch. That is, you don't need to insert a physical SIM card, like the one in your iPhone, to access the LTE and UMTS cellular bands.

Apple Watch Ultra only ships in a GPS + Cellular option.

Apple Watch now supports worldwide roaming.

Beginning with Apple Watch Series 5, emergency calling works in countries outside the United States (see Chapter 5).

NFC

NFC (near-field communication) is a short-range radio technology (like Bluetooth) that has numerous applications but is most commonly associated with mobile payments, similar to waving or tapping your iPhone on a contactless terminal at retail locations (or a compatible vending machine) to make a secure purchase. Apple Watch also uses NFC to make a digital handshake with the terminal to complete the transaction. Yep, it's all in the wrist. This feature is part of Apple Pay, Apple's mobile payment solution for secure cash- and cardless payments. Check out Chapter 10 for more on Apple Pay.

GPS

Except for the first Apple Watch (Series 1, from 2015), Apple Watch has an integrated GPS chip to identify its location on Earth down to a few meters of accuracy. Therefore, when coupled with mapping applications, GPS can help you see your location on a map, get directions from point A to point B, look for local businesses of interest, and more. GPS can also help with tracking fitness data when measuring steps won't help (such as in cycling). Along with the accelerometer (discussed next), built-in heart rate sensor, and Wi-Fi, Apple Watch's GPS can help measure distance traveled. Jog on over to Chapter 8 to learn more about the Activity and Workout apps.

TIP

Ever want to take a screen shot of something on your Apple Watch, such as an impressive day of physical activity? Press and hold the side button; then tap the Digital Crown. You'll hear a shutter button, the watch face will flash white, and the image will appear wirelessly in your iPhone's photo gallery. You may need to enable screen shots first by opening the Apple Watch app on an iPhone and tapping My Watch ➪ General ➪ Enable Screenshots.

Ultra Wideband

While there aren't a lot of use-cases for it yet, Apple added Ultra Wideband (UWB) technology to its Apple Watch Series 7 and Apple Watch Series 8. Similar to iPhones with UWB (beginning with iPhone 11 in 2019), this feature provides more precise location and spatial awareness. As you'll see in Chapter 10, you can unlock a compatible car without even having to hold the iPhone or Apple Watch near the door lock.

Accelerometer, gyroscope, barometric altimeter, and compass

Like other smartwatches and activity bands on the market, Apple Watch has an accelerometer that measures movement, whether you're lifting the watch to your face to turn on the screen; lowering your wrist to decline a call; or calculating fitness activities, including steps taken (like a 21st-century pedometer), total distance traveled, time spent exercising, and estimated calories burned. Beginning with Apple Watch Series 3 (2017), the watch also has an integrated barometric altimeter for measuring elevation (counting the steps you climb or descend) as well as calculating altitude for those who ski, hike, or climb mountains. Beginning with Apple Watch Series 6, the altimeter is always on, providing all-day, real-time elevation data on your wrist.

With the Apple Watch's accelerometer and gyroscope sensor, Apple Watch Series 4 (2018) and newer models can detect whether you've fallen, and you can initiate a call to emergency services (or dismiss the alert). If you're unresponsive after 60 seconds, Apple Watch automatically places the emergency call and sends a message with your location to your emergency contacts. *Note:* This feature is different from speaking to a live operator if you call, as you would with a service such as Philips Lifeline, but this feature is very handy to have on your wrist nonetheless (and you have no monthly fees to worry about).

A compass has been added to Apple Watch Series 5 and Series 6 models, always pointing you north inside apps like Maps, and there's a dedicated Compass app too. (Cue the song "Go West" by the Pet Shop Boys or the Village People first!)

Heart-rate sensor

A custom heart-rate sensor included with Apple Watch helps you in two ways.

» **Gauging your exercise intensity and tracking overall calorie burn:** (This data is an estimation based on info you input one time, such as your height, weight, gender, and age.) Apple Watch listens to your heartbeats per minute and shows you data on the screen — if and when you call for it.

» **Tracking your heart rate throughout the day:** Apple Watch can alert you if it detects unusually high or low heart rates — and yes, you can set the parameters if you want, even if you don't feel symptoms. Behind the watch are multiple sensors that measure your pulse through your skin. Going beyond fitness are fun applications, such as the one that lets you share your heartbeat with someone — felt on their Apple Watch — to show you're thinking about them. See Chapter 5 for how to share this information to your heart's content.

Electrocardiogram (ECG)

Beginning with Apple Watch Series 4 (except Apple Watch SE), all Apple Watch models include an electrical heart-rate sensor that can take an electrocardiogram (ECG) by using an ECG app; the built-in sensor and the electrodes are included in the Digital Crown button. You take an ECG reading by placing a finger on the Digital Crown while wearing Apple Watch; the reading is completed within 30 seconds. The ECG app tracks whether your heart is beating in a regular pattern or whether you have signs of atrial fibrillation — a clear indicator of serious health problems.

Blood-oxygen monitor (pulse oximeter)

Beginning with Apple Watch Series 6, your wearable device is capable of detecting blood-oxygen levels. When you tap the Apple Watch app, green, red, and infrared LEDs shine light into your wrist, and photodiodes measure the amount of light reflected back. Then advanced algorithms calculate the color of your blood, indicating the amount of oxygen present. Wow. Blood-oxygen levels between 95 and 100 percent are considered to be healthy — a lower percentage *could* indicate compromised heart, lung, or brain functionality — but be sure to consult your doctor. In fact, Apple stresses that the Blood Oxygen app is not intended for medical use; it's designed for general fitness and wellness purposes.

Ambient light sensor

Finally, Apple Watch has an ambient light sensor under the glass. This sensor samples the environmental light falling on the screen and automatically adjusts the brightness up or down to improve viewing comfort so that the screen isn't too dim or too bright, based on where you are. An ambient light sensor also helps regulate the power that the display uses, thus squeezing more battery life out of the watch.

In fact, beginning with Apple Watch Series 5, the Retina screen is always on — just dimmed — until you turn it toward your face to read the time or other info (except Apple Watch SE devices, which do not have this feature). In previous Apple Watch models, the screen goes black when you're not looking at it, and it takes a split second to wake up and turn on when you turn it toward you.

Crash detection

You likely know about fall detection (see Chapter 8) but now Apple Watch offers crash detection for drivers and passengers.

Apple Watch Series 8, Apple Watch SE (2022 model), and Apple Watch Ultra have added crash detection, which relies on an advanced sensor-fusion algorithm from Apple that leverages a new, more powerful gyroscope and accelerometer on Apple Watch. Tested in crash test labs with common passenger cars — including simulated head-on, rear-end, side-impact accidents, and rollovers — your Apple Watch can detect a car crash, call emergency services (and your chosen contacts), and share your location data. In addition to motion data, crash detection uses the barometer, GPS, and the microphone on iPhone as inputs to detect the unique patterns that indicate whether a severe crash has taken place (such as if the air bag is deployed). Neat!

Temperature sensors

Apple Watch Series 8 and Apple Watch Ultra offer temperature sensing, with a two-sensor design: one sensor on the back of the watch, nearest the skin, and another just under the display (reducing bias from the outside environment). Nighttime wrist temperature can be a good indicator of overall body temperature, but Apple says the primary focus is on women's health, such as retrospective ovulation estimates. Knowing when ovulation has occurred can be helpful for family planning. These estimates can be viewed in the Health app. Temperature sensing also enables improved period predictions. See Figure 1-14.

FIGURE 1-14:
Apple Watch
Users who
utilize the new
temperature-
sensing
capabilities in
Apple Watch
Series 8 or
Apple Watch
Ultra can
receive
retrospective
ovulation
estimates
and improved
period
predictions,
which can
be helpful
for family
planning.

Tapping with Apple Watch's Haptic Feedback

You can tap Apple Watch's screen, and guess what? It can tap you too.

Like videogame controllers that vibrate when your soldier gets shot or some smartphones and tablets that buzz slightly when you tap a letter on the virtual keyboard, Apple Watch employs haptic technology to apply light force to your skin to alert you to relevant information. Apple's *Taptic Engine* is a linear actuator inside the watch that produces discreet haptic feedback.

Consider this slight vibration to be a third sense (touch), along with sight and sound, to give you information. The physical sensation of a tap tells you something, such as a warning that an important meeting is about to start, without your even having to look at your wrist. This feature can also be a silent alarm clock to wake you up in the morning instead of bothering your significant other. Or it can transmit the feeling of your loved one's heartbeat even though they may be miles way.

What's more, Apple Watch can tap different patterns based on who's reaching out to you (such as two taps for your spouse and three taps for your boss). Or perhaps the haptic pattern tells you what the information is: one tap for the time on the hour, four taps for a calendar appointment, and so on. Neat, huh?

In the near future, Apple Watch's haptic feedback may let you know about important health information, perhaps working in conjunction with sensors. Imagine if someone living with diabetes could get a haptic tap to tell them it's time to use insulin, based on their body's blood-sugar levels.

Chapter **2**

Time Out: Setting Up Your Apple Watch

W ell, you did it. You're now the proud owner of an Apple Watch. Or perhaps you received one as a gift, and you're more intimidated than proud (I get it).

Regardless, I'm thrilled you picked up (or downloaded) this book to help you get the most from your wearable companion. You're gonna love your new gadget.

This chapter helps you set up Apple Watch for the first time and covers how to charge it, take care of it, and get to know some of the basics.

Setting Up Apple Watch

If you're a fan of those unboxing videos on YouTube, you've watched a few gadget geeks excitedly open a box for the first time and expose all the goodies inside. But if you don't have the time or desire for that drama, let me tell you what you can expect if you haven't received your Apple Watch just yet or what you should find in your box if you have.

In your Apple Watch box, you should find:

>> Apple Watch

>> Band (what kind will vary by model)

>> Magnetic-dock charger cable (USB cable)

>> Small booklet with setup and maintenance tips

Be sure to keep the box and all the things inside just in case you need to return or exchange the watch.

Wait — where's that little cube thingy that plugs the cable into the wall to charge up Apple Watch? As Apple announced, it has stopped shipping its power adapter with Apple Watch and iPhone. If you're wondering why, the company says it's to help the environment — both in the materials used to make this doohickey and potential electronic waste (e-waste) when disposing of it — and we already have a couple in a box or drawer somewhere, in all likelihood. Others say that Apple is trying to save money. Either way, you can still plug the watch into a powered USB-A port on a computer or use another 5-watt power adapter cube you have lying around.

THE LIMITS OF WATER RESISTANCE

Beginning with Apple Watch Series 2, you can wear all models in the shower, outside in the rain, while sweating during a workout, or while swimming! But be aware that Apple Watch is meant for shallow-water activities: Apple cautions that you shouldn't wear it for scuba diving, water skiing, or "other activities involving high-velocity water or submersion below shallow depth." Also avoid soap and shampoo, if you can, because they could affect the watch's performance. Near the end of this chapter, I share some obvious (and not so obvious) ways to take better care of your Apple Watch.

Like many consumer electronics you buy today, the watch might already be charged when you get it, but it's always a good idea to plug it into a computer or the wall to give it a full boost before using it for the first time. This way, you won't run out of juice while playing around with the watch's settings, learning the mechanics, and so on.

Pairing Apple Watch with the iPhone

After you ensure that your smartwatch is charged, follow these steps:

1. **Turn on your Apple Watch by pressing and holding the side button.**

   This button is the one that's flush with the watch, not the Digital Crown button that's sticking out. When you press it, the Apple logo appears in the center of the screen, which is a good sign!

2. **Tap the Apple Watch app on your iPhone.**

   This app is a black icon that simply says Watch. If you don't see the app on your phone's Home screen, swipe left or right to look for it.

TECHNICAL STUFF

   You need an iPhone 5 or newer and the 8.2 iOS (or newer) operating system installed to use Apple Watch for Series 1 and 2, and an iPhone 6 and iOS 12 or later for Series 3, Series 4, and Series 5. For Apple Watch Series 6 and Apple Watch SE, you need to have iOS 15 or later. To double-check what you have, tap Settings ⇨ General ⇨ About, and look for Version. Your phone also notifies you about any available updates.

   The latest operating system, watchOS 9, requires iPhone 8 or later with iOS 16 or later and one of the following Apple Watch models:

   - Apple Watch Series 4
   - Apple Watch Series 5
   - Apple Watch SE
   - Apple Watch Series 6
   - Apple Watch Series 7
   - Apple Watch Series 8
   - Apple Watch Ultra

   When you bring the Apple Watch close to your iPhone, you should see the words *Use your iPhone to set up this Apple Watch* on your iPhone screen. If you don't, open the Apple Watch app on the iPhone and then tap Start Pairing.

3. **Tap Continue.**

   Keep your Apple Watch and iPhone close together until you complete these next steps.

4. **Follow the prompts.**

   The prompts ask you to hold Apple Watch up to the iPhone's camera. Then you can align the watch's face within the viewfinder in the center of the screen, which should do the trick. See Figures 2-1 and 2-2 for a look at setting up Apple Watch for the first time.

   If that process doesn't work, tap the Pair Apple Watch Manually option, in orange, at the bottom of the app. You're prompted to tap the "i" (information) app on your Apple Watch to view its name and then tap the corresponding name listed in the app. If it's not listed, make sure that your wireless connection is enabled; then swipe up from the bottom of the screen and tap the icons for Bluetooth and Wi-Fi so they're highlighted and not grayed out.

TIP

   If you're prompted to set up a cellular connection, read the later section "Setting up cellular connectivity on Apple Watch."

   You're almost done!

5. **If this Apple Watch is your first, tap Set Up Apple Watch.**

   Otherwise, choose a backup. If prompted, update your Apple Watch to the latest version of watchOS, the name of Apple's operating system that powers this wearable.

6. **Read the terms and conditions and tap Agree (twice).**

   If asked, enter your Apple ID password. If you aren't asked, you can sign in later from the Apple Watch app (General ⇨ Apple ID). Some features that require a cellular phone number won't work on cellular models of Apple Watch unless you sign into iCloud.

7. **Choose a text size for the Apple Watch.**

   You may want to make the font larger, for example, to be easier to read. Who needs the crow's feet? (Am I right?)

Your Apple Watch shows you which settings it shares with your iPhone. If you've enabled Find My iPhone, Location Services, Wi-Fi Calling, and Diagnostics on your iPhone, for example, these settings automatically turn on for your Apple Watch. You can select other settings too, such as Siri (your personal voice-activated assistant) and Route Tracking.

FIGURE 2-1:
The Apple
Watch app
asks you to
pair your
Apple Watch.
Easy-peasy.

When the pairing is successful, you can adjust additional watch settings from
within the app by tapping My Watch in the bottom-left section of the screen (see
Figure 2-3). Take some time to familiarize yourself with this great app; I spend a
lot of time on it for this book, too.

FIGURE 2-2:
Match up the
Apple Watch
inside the
outline on your
iPhone screen.

In fact, the three main sections to the Apple Watch app are listed at the bottom of the screen:

>> **My Watch:** This section is your main section, which lets you customize your watch face (Complications), enable or disable features, tweak settings, and enable notifications per app. It also has a Search window if you want to type a keyword.

>> **Face Gallery:** Tap this section to view and change your Apple Watch clock faces. (See Chapter 4 for all your options.)

>> **Discover:** This section offers information on using your Apple Watch (such as customizing it), as well as an Explore Watch Apps area at the bottom of the screen where you can download and install third-party apps on your device (yes, an app store).

Keep in mind that you don't need to turn Apple Watch on or off. Simply raise your wrist, and the screen turns on — thanks to its internal accelerometer (motion sensor) — and lower your arm to turn it off. It's that easy. Or if you have an Apple Watch Series 6, 7, or 8, or Apple Watch Ultra model, the screen never goes dark; it stays on but dims itself until you look at it!

TECHNICAL STUFF

How does Apple Watch's beautiful Retina display stay on all the time without killing the battery? At the risk of geeking out, I can tell you that it uses an LTPO (low-temperature polysilicon and oxide) display that drops the screen's refresh rate from 60 Hz to a "power-sipping" 1 Hz when the watch is inactive (that is, when you're not looking at it). A low-power driver, ambient light sensor, and efficient power management software also work together to keep your watch going for up to 18 hours between charges (or up to 36 hours in Low Power mode, plus Apple Watch Ultra can last up to 36 hours or up to 60 hours in Low Power mode). Just touch the screen or point it toward your face for full brightness. Cool, eh? Apple made several improvements with the larger Apple Watch Series 7 too, making the always-on display even brighter without affecting battery life. See Figure 2-4.

FIGURE 2-3:
Although you may be eager to play around with your new Apple Watch, spend some time familiarizing yourself with the Apple Watch app on the iPhone as well.

FIGURE 2-4:
Since the debut of the Apple Watch Series 5, all models (except Apple Watch SE) have used a new kind of screen that never turns off 100 percent. Instead, the screen dims when you're not looking at it.

Choosing a passcode for Apple Watch

Although doing so isn't mandatory, those who use Apple Watch can set up a personal identification number (PIN) in case the watch is lost or stolen. Some people call a PIN a *passcode*, but it's the same thing.

TECHNICAL STUFF

A PIN is mandatory when you use Apple Pay to buy goods or services through your smartwatch. See Chapter 10 for more on Apple Pay.

Just like with an iPhone, iPad, or iPod touch, if the Apple Watch's PIN option is activated, you have to enter the four-digit numeric code (or you can choose a longer one) every time you put the watch back on. Apple Pay requires skin contact and the PIN. See Figure 2-5 for details on setting up a PIN on Apple Watch.

FIGURE 2-5: A passcode is optional for Apple Watch unless you use Apple Pay.

Wait — how does Apple Watch "know" whether you're taking the watch off or putting it on? The smartwatch's internal sensors touch your skin. I cover this topic in Chapter 10, where I also go over what to do if you lose your Apple Watch. (*Hint:* As with a lost iPhone, you can go to www.icloud.com, log in with your details, and remove your credit and debit cards in case the watch falls into the wrong hands.)

Bottom line: You may want to set up a passcode on your Apple Watch even if you don't activate Apple Pay. And of course, you should also have a PIN set up on your iPhone.

After you set up a passcode on Apple Watch, you may need to wait a little while for your devices to sync. How long really depends on how much data you have, but the wait should be only a few minutes at longest. Be sure to keep your devices close together until you hear a chime and feel a gentle tap from your Apple Watch to confirm that they have synced. Then press the Digital Crown button to see your apps.

Setting up cellular connectivity on Apple Watch

If you're reading this section, you've got one of the Apple Watch models that supports cellular connectivity. This feature is super-exciting because it lets you use your watch for calls, texts, real-time notifications, streaming music, and more, even when you're away from your iPhone. Noncellular models require Apple Watch to be near an iPhone for most connectivity tasks, such as placing calls, sending messages, accessing online content, and receiving notifications.

Apple isn't selling certain Apple Watch versions any longer — including Apple Watch Series 4, which also has cellular connectivity on some models — so don't worry if you have one of those versions. You can still follow along here with the following tips.

To use your Apple Watch's cellular functions, you need

>> An Apple Watch that supports GPS + Cellular, such as Apple Watch Series 3, Apple Watch Series 6, Apple Watch Series 7, Apple Watch Series 8, Apple Watch SE, or Apple Watch Ultra (GPS + Cellular only).

>> An iPhone 6 or later with the latest iOS version.

>> An eligible cellular service plan with a supported carrier. (The iPhone and Apple Watch must use the same carrier.)

Setting up your GPS + Cellular watch is easy. You can follow along with the onscreen steps (shown in Figure 2-6). Alternatively, you can set up cellular later from the Apple Watch app. The following steps walk you through connection and use of the watch afterward:

1. **When you first set up your new Apple Watch, follow the prompts to activate cellular.**

   You see the start screen in the image on the left side of Figure 2-6.

   When the feature is activated, note the color of the Cellular button on your Apple Watch:

   - *Green:* Indicates that you're using cellular connectivity. The green dots over the button show the signal strength (the more dots, the stronger and better the signal). See the middle image in Figure 2-6.

   - *White:* As the right image in Figure 2-6 shows, the once-green button turns white when your cellular plan is active but your watch is connected to your iPhone via Bluetooth or Wi-Fi.

2. **On your iPhone, open the Apple Watch app, tap the My Watch tab in the bottom-left corner of the screen, and then tap the green icon that says Cellular.**

3. **Follow the onscreen instructions.**

Now, going forward, your Apple Watch automatically reverts to the most power-efficient wireless available, whether it's a nearby iPhone via Bluetooth, a Wi-Fi network, or cellular connectivity. When your watch connects to cellular, it uses LTE/4G networks. (Nope, Apple Watch doesn't support 5G networks yet.) If LTE isn't available, your watch attempts to connect to UMTS, should your carrier support it.

Some tips and tricks to remember:

>> **To check the signal strength of your carrier network:** Swipe up from the bottom of the Apple Watch screen to see the Control Center interface, or select the Explorer watch face to see your signal strength at a glance.

>> **To check your cellular data use:** Open the Apple Watch app on your iPhone, tap the My Watch tab, select Cellular, and scroll to the Cellular Data Usage section.

>> **To enable or disable your cellular connection:** Simply swipe up from the watch face to open Control Center, tap the green icon with the white wireless symbol inside, and turn Cellular On or Off.

Family Setup: Setting up Apple Watch for family members

As you might have heard from friends or read in this book, Apple Watch previously required you to have an iPhone to set up and use the watch (even with the cellular model). But this hasn't been the case since the introduction of watchOS 7 in the Fall of 2020. (We are up to watchOS 9 as I write this book.)

With a new feature called Family Setup, it's possible for family members to use Apple Watch's features — such as sending and receiving messages and calls and sharing their location with you — even if they don't own an iPhone. After you set up a watch for a family member, such as a child, you can use your iPhone to manage some of the watch's capabilities.

To get going with Family Setup, you need

» An Apple Watch Series 4 or later, GPS + Cellular model, or Apple Watch SE with cellular, or Apple Watch Ultra. Apple says a cellular plan (typically, $10 per month) isn't required to set up an Apple Watch for a family member, but it's necessary for some features.

» watchOS 9 requires iPhone 8 or later, running iOS 16.

» Your Apple ID and one for the family member who will use the Apple Watch. Note that two-factor authentication must be turned on. (In *two-factor authentication,* you need not only a password to log in, but also a one-time code sent to one of your Apple devices, which you also need to type to confirm that you're really you.)

Ready to get going? Follow these steps to use Family Setup as the parent/guardian:

1. **Put the Apple Watch on your wrist and turn it on by pressing and holding the side button.**

    The Apple logo appears. If your Apple Watch isn't new, perform a factory reset on the watch (Settings ➪ General ➪ Reset ➪ Erase All Content and Settings).

2. **Hold the watch close to your iPhone.**

    Your phone's screen displays the message *Use your iPhone to set up this Apple Watch*.

3. **Tap Continue.**

4. **Pair the watch with your iPhone by following the prompts to hold the iPhone over the Apple Watch screen (so that your iPhone's camera can see the animation).**

    See Figure 2-7.

5. **Tap Set Up Apple Watch, and after you agree to the terms and conditions, choose a text size for the Apple Watch and a passcode.**

6. **Pick a family member who will use this Apple Watch.**

   You should see the family member's name, face, and age. If not, tap Add New Family Member and then enter their Apple ID and password. If you like, enable Ask to Buy if you want to give permission for any app downloads or purchases made on the Apple Watch.

7. **Set up cellular and Wi-Fi.**

   During this step, you can add your Apple Watch to your mobile-phone plan if your cellular provider supports it, or you might be able to use a different carrier. If that's the case, leave this part for later. But be sure to choose whether to share your current Wi-Fi network with the Apple Watch.

8. **Evaluate other features.**

   On the next few screens, you can enable or disable various Apple Watch features, including Location Services (for the Find People app), Siri, Apple Cash Family, Messages in iCloud, Health Data, Emergency SOS, Emergency Contacts, Medical ID, Activity, Workout Route Tracking, and Photos.

9. **Set up shared contacts and Schooltime.**

   You're asked to set up the approved contacts available on Apple Watch to call or message (such as family members, friends, or a trusted neighbor). To do so, enable Contacts in iCloud (on the iPhone, tap Settings ⇨ [your name] ⇨ iCloud, make sure that Contacts is turned on, and select people.

TIP

Apple has set some restrictions on kids using Apple Watch. For one, high and low heart-rate notifications are available only for users 13 and older, and fall detection is available for users 18 and older (see Chapter 8).

Other health and wellness features that aren't supported in Family Setup include irregular-heart-rhythm notifications, ECG, Cycle Tracking, Sleep, Blood Oxygen, Podcasts, Remote, News, Home, and Shortcuts.

Apple Cash Family is available to users under the age of 18, allowing them to make purchases (and send and receive money) in the Messages app via Apple Pay (United States only). See Chapter 10 for more information. But Apple Pay via a credit, debit, or transit card in the Wallet app isn't available through Family Setup.

Added in 2022 with watchOS9, Family Setup now lets kids access podcasts (with your approval), control HomePod speakers and smart home accessories, and share some wallet cards, like virtual hotel keys, so kids can get into the room.

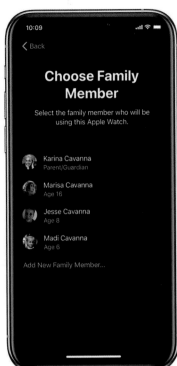

FIGURE 2-7:
Apple Watch
lets you set
up a watch for
someone in
the family who
doesn't have
an iPhone.

Enabling Schooltime on a family member's Apple Watch

Introduced in watchOS 7, Schooltime — as the name suggests — was designed to limit the use of Apple Watch during specific days of the week and hours of the day. Figure 2-8 shows the process of setting up Schooltime in the iPhone's Watch app (and on the watch itself).

To set a Schooltime schedule, follow these steps:

1. **Open the Watch app on the iPhone, tap All Watches, select the name of your child's watch, and tap Done.**

2. **Select Schooltime and tap Edit Schedule.**

3. **Choose the days and times when you want Schooltime to be activated (such as weekdays between 8 a.m. and 3 p.m.).**

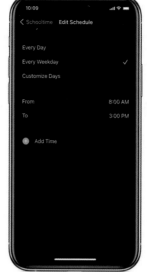

FIGURE 2-8:
The Schooltime feature lets you set specific dates and/or times you don't want the child to access Apple Watch.

Additional notes:

>> If you want to set up more than one schedule in a day (such as a morning routine and an afternoon one), open the Watch app on your iPhone, tap the information ("i") button, and then tap Schooltime and Edit Schedule.

>> If necessary, the student could exit Schooltime any time by turning the Digital Crown, and then tapping Exit to confirm. The starting and ending times of all temporary Schooltime exits are reported in the Watch app on the iPhone, and parents can see them by opening the Watch app, tapping the child's Apple Watch (listed in All Watches), and then selecting Schooltime.

Monitoring the Apple Watch battery

One of the biggest challenges of such small technology that's always on? Battery life. Thus, Apple challenged its engineers to squeeze all-day performance out of Apple Watch. And they succeeded.

Okay, so *all-day performance* is a little vague, but as noted in Chapter 1, Apple says it amounts to "up to 18 hours" — based on Apple's testing. *Note:* This is for Apple Watch Series 7, Apple Watch Series 8, and Apple Watch SE. Apple Watch Series 7 and Apple Watch Series 8 models have even-brighter always-on Retina displays. In Low Power mode, which disables some features, battery life could top 36 hours. Apple Watch Ultra's battery lasts up to 36 hours, or up to 60 hours in Low Power mode.

Apple Watch includes a magnetic charging cable. One end snaps onto the back of the smartwatch and is secured by a magnetic connection — not unlike Apple's MagSafe charger for MacBook laptops — and the other end of the cable can plug into a computer's powered USB port (Mac or PC) or into a traditional electrical socket (with an adapter on the end, which you need to provide on your own).

As you might've noticed — in this text or after looking in the box Apple Watch came in — the USB cable doesn't plug into Apple Watch anywhere. Unlike some other smartwatches, this one has no port to uncover. Instead, the circular puck magnetically affixes to the underside of the watch, where the heart-rate sensors are, and powers up the watch through induction technology.

The magnetic charger makes it easy to juice up Apple Watch because you don't have to open any ports on the watch to plug in a cable. Just attach it to the back of your Apple Watch, snap it into place, and you're good to go. See Figure 2-9 for a look at the unique Apple Watch charger.

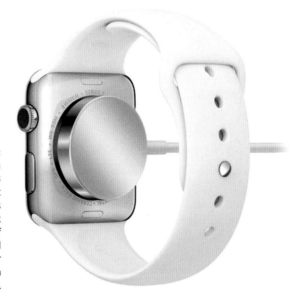

FIGURE 2-9:
Apple Watch has no USB ports. Just connect this magnetic puck to the back of the watch and plug the other end into a computer.

Broken down, the 18-hour battery life (on Apple Watch Series 7, Series 8, and SE models) includes the following:

>> Ninety time checks (4 seconds long apiece)

>> Receipt of 90 notifications

>> Forty-five minutes of app use

>> A 60-minute workout with music playback from Apple Watch via Bluetooth

Newer and larger Apple Watch models typically experience longer battery life. The company also cautions that "battery life varies by use, configuration, and many other factors; actual results will vary."

BATTERY-RELATED ACCESSORIES FOR APPLE WATCH

If you're on the go and don't want to worry about plugging the watch in somewhere to juice up, several powercentric products can help.

The MIPOW Portable Apple Watch Charger ($89) is a pocket-size wireless charger for the Apple Watch, featuring a 10,000-milliamp (mAh) battery and integrated Lightning cable for charging up an iPhone and iPad too, if desired. MIPOW says that this Apple MFi Certified solution can charge up an Apple Watch to about 50 percent in just 30 minutes.

(continued)

(continued)

Available in several colors, the portable product powers up the watch through magnetic induction, and a USB port is used to power up the MIPOW battery for when you need it next.

A similar option is the Kanex GoPower Watch Plus ($69), a portable 5,200 mAh battery with an integrated magnetic inductive charger for Apple Watch. Like the MIPOW product, this battery can charge Apple Watch and an iPhone simultaneously. Kanex says you can charge an Apple Watch up to eight times before the GoPower Watch Plus needs recharging. The unit holds the magnetic watch charger in a cradle, and you can feed the charging cable through to the back of the stand to reduce clutter.

If you're curious about specific tasks, Apple breaks down battery performance even further (based on its testing):

>> **Talk-time test:** Up to 3 hours. Apple Watch was paired with an iPhone during the call.

- » **Audio playback test:** Up to 6.5 hours (when paired via Bluetooth with an iPhone).

- » **Workout test:** Up to 6.5 hours. Apple Watch was paired with an iPhone and had a workout session active and the heart-rate sensor turned on.

As Apple suggests, you likely want to charge the watch at the end of each day to get it ready for the following one — unless you want to use the Sleep Tracking feature (see Chapter 8). Should you want it, more information about battery life is available at www.apple.com/watch/battery.

Apple says that its Apple Watch Series 7 models charge up to 33 percent faster than their predecessors. Cool.

Understanding the Home Screen

Although you can customize what app you see first, by default, the clock is what appears when you look at the watch. If you press the Digital Crown button, however, you can access your Home screen to see all the apps installed on the watch.

Apple Watch's Home screen, shown in Figure 2-10, is similar to other iOS devices — namely, the iPhone, iPad, and iPod touch. Basically, you'll see a bunch of icons that launch an app when you tap one.

The Home-screen apps include Apple's own applications, such as a speech bubble for Messaging or a cloud with a sun peeking behind it for Weather. In addition, you'll see any third-party apps you choose to install and transfer to the watch from your iPhone. These apps might include a social media feed, such as Twitter, news information provided by CNN, or a game such as Trivia Crack. You also can see many of the native apps for Apple Watch. (Refer to Table 1-1 in Chapter 1.)

But instead of the neatly arranged rows and columns of app icons you see on an iPhone or iPad, Apple Watch apps are arranged like bubbles in different sizes that move as you press and slide your finger around the Home screen. The apps grow larger when the icons get closer to the center of the Apple Watch screen, making them easier to tap and launch.

FIGURE 2-10: Although the icons are bubbles instead of rounded rectangles (on the iPhone and iPad), your Home-screen apps should still look familiar to you.

If you twist the Digital Crown button while you're viewing the Home screen, you can zoom in and out of all the apps you have installed on the watch.

TIP

If you zoom enough on one app, Apple Watch launches it for you! Also, while you're in the Clock app, you can simply press the Digital Crown button to return to the Home screen.

You can choose which apps are installed on Apple Watch via the Apple Watch app on your iPhone (preinstalled in iOS 8.2 and later versions). An Apple Watch App Store is also built into the Apple Watch app for iPhone to allow you to download and manage your apps there (or you can download apps to the Apple Watch directly, if you prefer). See Chapter 11 for directions on installing third-party apps.

Figure 2-11 shows the Apple Watch app on an iPhone.

REMEMBER

If you hold the Digital Crown button down for more than a second, it launches Siri, your personal voice-activated assistant (or simply say "Hey, Siri" into your watch). Only a quick tap on the Digital Crown button is necessary to open your Home screen. See Chapter 7 for more on using Siri on your Apple Watch.

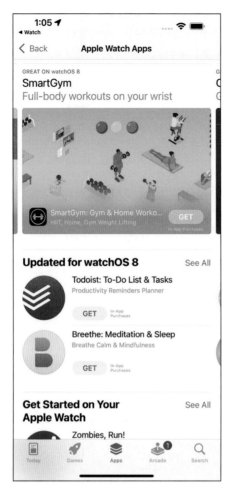

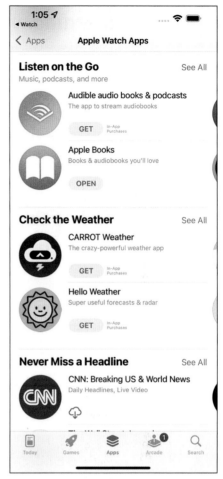

FIGURE 2-11:
Tap App Store
inside the
Apple Watch
app on your
iPhone to
download new
apps (left).
Featured apps
appear divided
by theme,
games, news,
and other
categories
(right).

Maintaining Your Apple Watch

Safeguarding your Apple Watch investment is probably one of the wisest actions you should take. After all, unless you received it as a gift (lucky you), you probably spent many hundreds on your Apple Watch, so it's a good idea to proactively protect it from damage.

Because your smartwatch is wearable, at least you don't have to worry so much about dropping it — as you would a smartphone, tablet, or laptop — but you should still take steps to ensure that your Apple Watch runs smoothly for many years.

Avoiding water (original Apple Watch)

The first tip is to be mindful of water if you own the original Apple Watch (Series 1), which debuted in 2015. This particular first-generation Apple Watch isn't waterproof. It might be splash-resistant, but you shouldn't fully submerge your wrist in water while wearing it. Keep your wrist away from the faucet when washing your hands or doing the dishes, and don't forget to take it off when you hop into the shower, climb into a bath, or step into any body of water. Your existing watch might be waterproof — and old habits die hard, as they say — so don't forget to take off Apple Watch before you do anything involving water. You really don't need to see who's texting you while you're soaking in a hot tub.

Getting caught in the rain or sweating profusely during a run are okay, says Apple, but err on the side of caution to be extra-safe, and leave Apple Watch behind before you go boating. Jogging with Apple Watch on a cloudy day? Don't worry about taking an umbrella, because rain won't harm your gadget.

If you own Apple Watch Series 2 or newer, however, it's indeed water-resistant, and your device goes into Water Lock mode when it senses moisture! But here's something you should consider after a swim: Turn the Digital Crown button to unlock the screen, and you may hear sounds emanating from the Apple Watch and feel some water on your wrist from the teeny speaker. To clear water from your Apple Watch manually, swipe up on the bottom of the watch face to open Control Center, tap Water Lock, and twist the Digital Crown button to unlock the screen.

Apple says that its latest series, Apple Watch Ultra, is the most durable to date. Rated IP6x dust-resistant and tested to MIL-STD 810H3 specs, the titanium-based Apple Watch Ultra is corrosion-resistant and has water resistance to 100 meters.

Avoiding extreme temps

Be cautious when using your Apple Watch in extreme temperatures. This consideration may be relevant based on where you live or work. Are you reading this book while lying by the pool in Acapulco or Maui? Or perhaps you work in the Athabascan oil sands up in northeastern Alberta, Canada.

Apple hasn't said what the optimum environmental requirements are for Apple Watch Series 7, 8, or SE, but assuming for a moment they're similar to those of its iPhone, iPod touch, iPad, and MacBook laptops, you should keep the following parameters in mind:

- » **Operating temperature:** 32° to 95° F (0° to 35° C)

- » **Nonoperating temperature:** –4° to 113° F (–20° to 45° C)

- » **Relative humidity:** 5 percent to 95 percent noncondensing

To be sure, check Apple's website or ask someone who works at an Apple Store.

Doubting its durability

Although Apple advertises how strong its smartwatch is — what with its solid materials, reinforced glass, and durable wristbands — try to remember that you've got a sophisticated computer on your left or right wrist.

Remove the watch whenever you're doing something that could potentially damage it, such as tossing a baseball with a child in a park or working under your car in a garage. Accidents happen, sure, but avoid the chance of impact to the area or the odds of harmful fluids dripping onto the watch.

Also consider the bumps and knocks of everyday life: slamming a school locker shut, wrestling with your dog, or even putting away the dishes in drawers and cupboards. Don't be afraid to wear and use Apple Watch, but remember that it's not immune to damage.

Considering a bumper

Just as other Apple products have countless accessories, you can bet that Apple Watch is starting to see its fair share of optional add-ons to help you get more from the wearable — including protective bumper cases. One of the most affordable is Modal's Apple Watch Bumper watch case, at $7 to $10.

As you might suspect (or glean from the image in Figure 2-12), the rubber Bumper has been designed to snugly fit and protect your Apple Watch and make it actionproof — hence, the name of the company.

Although the Bumper wraps around the body of the watch, you still have access to the Digital Crown button and the side button. The case also allows for full access to the screen, backside sensors on your skin, and unobstructed use of the microphone and speaker.

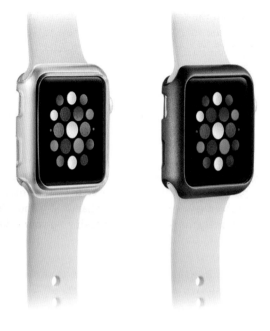

Apple Watch sold separately.

Using Apple Watch responsibly

Even though much of the advice I dispense is common sense, you'd be surprised just how often it's forgotten or ignored. Thus, the following sections offer some suggestions for being smart about your smartwatch.

Watch the road, not your wrist

Just as you shouldn't be distracted by other technology while you're behind the wheel — holding up a smartphone, glancing at a tablet, or fiddling with a GPS navigation device — it's critical that you resist accessing Apple Watch while driving a five-ton vehicle. As the late, great Jim Morrison once famously sang: "Keep your eyes on the road — your hands upon the wheel."

"But I don't have to hold a smartwatch," you say.

True, but you can still tap the screen, press one of its two buttons, look down to read something, or hold the small speaker up to your ear to listen to Siri's voice — all of which could temporarily, and even fatally, distract you when you should be concentrating on the task at hand.

We all know this, of course, but look to your left or right while you're stopped at a stoplight, and no doubt you'll see someone using a tech gadget while in the driver's seat.

Using your watch while driving may be temping, but wait. The more technology we have at our fingertips, the more likely we are to use it wherever and whenever. Even using hands-free technology has been proved to distract drivers, so although you may be looking at the road and keeping your hands at 9 and 3 (or 10 and 2?) as your driving instructor likely taught you, not focusing on your driving exclusively could be an issue.

I'm not here to lecture, of course, but if this section reminds you about the often-overlooked dangers of distracted driving, it did its job!

Watch out for sidewalks too

Anyone who's spent time on YouTube might've seen some humorous videos of people walking down the sidewalk, oblivious to the world around them because they're staring at their smartphone, and as a result, they walk into walls or people, trip on a curb, or even fall into a manhole.

Funny? Sure. These clips prove that not all of us can multitask as well as we think, especially because we can't be looking at our phone and seeing what's in front of us at the same time.

But what you might not see on YouTube are distracted pedestrians walking into or across a road and getting hit by a vehicle. Those situations aren't so humorous, although you may be tempted to nominate these unfortunate souls for a Darwin Award, a silly online commemoration of people who perish in ridiculous ways to protect our gene pool as a nod to Charles Darwin's evolutionary theory.

One tiny little mistake in judgment, and you could be seriously injured or killed — or force a car to swerve out of the way and injure passengers or other pedestrians. It's happened.

Therefore, although Apple Watch was meant to be worn and used, be mindful of your surroundings — even when you're on foot. If you want to look at, talk to, or hear your wrist-mounted companion, stop walking first.

Remember netiquette — even when not on the Internet

Have you ever been to a restaurant and looked around at the other tables? You might have noticed something peculiar over the past few years: People are looking at their mobile devices more than they are at the people they're with.

Perhaps this phenomenon is an unfortunate sign of the times: a date night disturbed by someone checking the score of their favorite team, or a group of friends who'd rather advertise where they are to people they're not with than appreciate those they're sitting beside.

Will the same thing happen with smartwatches? Or maybe wearable technology will be more discreet than a smartphone, because you can casually glance at your wrist while sipping a drink instead of navigating through menus on a 6.5-inch handheld device.

For the sake of humanity in the digital age, we should hope that technology helps rather than hinders human interaction, whether it involves a couple enjoying a quiet dinner in a restaurant, kids sitting in a classroom, or business associates collaborating on a project. Technology has a time and a place, and smartwatches are technology.

Reduce the likelihood of theft

Another take on discretion: Because many people can't afford an Apple Watch, you might think twice about flaunting it on your wrist. I've had many emails from readers whose tech items were stolen — even brazenly, in public, and during the day — and the victims are left feeling angry and violated, as you might expect.

Many articles have been written about the dangers of tweens and teenagers showing off their new gadgets, such as smartphones and expensive headphones, so try to be a little more discreet with your Apple Watch. The last thing you want to do is tempt fate and lose your smartwatch forever — or, worse, put your life at risk over a mere gadget.

Also, although Apple Watch doesn't hold a lot of data (most of it is stored on your iPhone), you still don't want your personal property falling into the wrong hands.

This discussion may seem a little preachy, but I simply want to get some of these obvious — or perhaps not-so-obvious — safety, privacy, and common-sense tips out of my system, because I see the unspoken rules broken all the time.

Taking Advantage of Accessibility Features

Although I cover some of this in Chapter 11's discussion of customizing your Apple Watch experience, this section looks at Apple Watch's accessibility options, which refer to features designed for those with visual, aural, or other impairments, such as physical and motor skills.

Not unlike the myriad options in iOS (for iPhones), iPadOS (for iPads) and macOS (for Mac), Apple Watch's watchOS 9 comes loaded with accessibility options that you can enable and use, if needed. (And if you're gifting Apple Watch to a loved one, and they could benefit from one or more of these accessibility aids, be sure to let them know!)

How to Preview and Enable Accessibility Options on Apple Watch

There are two ways to open Accessibility settings:

>> On your Apple Watch, open Settings, then tap Accessibility.

>> On your iPhone, open the Apple Watch app, tap the My Watch tab, and then tap Accessibility.

Because you have a larger display in which to work with and a few more options to select, I'd recommend the latter option: Use your iPhone to select and enable accessibility features of Apple Watch, and it will be immediately, wirelessly, and magically sent over to your wrist-mounted device. (See Figure 2-13.)

FIGURE 2-13:
Whether you
want to enable
or tweak
settings on
Apple Watch
or the Apple
Watch app on
iPhone, you
have several
ways to add
accessibility
features.

Understanding Apple Watch's Accessibility Options

Similar to those available for other Apple gear, the accessibility features for Apple Watch fall into one of three categories: Vision, Hearing, and Physical/Motor Skills.

Here's a look at what's available today — divided into each of the three kinds of aids offered.

Vision

Here are your options for vision–related accessibility features on Apple Watch:

>> **VoiceOver:** VoiceOver is a built-in screen reader that audibly tells you what's seen on your Apple Watch, in one of 37 supported languages. All of Apple Watch's native apps, such as Mail, Calendar, Maps, and Messages, are compatible.

To turn on VoiceOver on your Apple Watch during initial setup, just triple-click the Digital Crown.

>> **Braille:** Apple Watch supports many international braille tables and refreshable braille displays. You can connect a Bluetooth wireless braille display to read VoiceOver output, including contracted and uncontracted braille. When you edit text, the braille display shows the text in context, and your edits are automatically converted between braille and printed text.

To set this up on Apple Watch, go to Settings ⇨ Accessibility ⇨ VoiceOver ⇨ Braille, then choose the display and tweak settings, if desired.

>> **Use Screen Curtain with VoiceOver:** If you use VoiceOver, you can turn on Screen Curtain for added privacy and security when you need it (such as when you're checking your bank balance), because it turns your display off yet keeps your device and VoiceOver navigation active.

>> **Zoom:** This is a built-in magnifier on Apple Watch. You can use the Digital Crown to move across the screen by rows or use two fingers to move around the screen. With magnification adjustable up to 15 times the native size, Zoom provides solutions for a range of vision challenges. When Zoom is enabled, simply double-tap your Apple Watch screen with two fingers to zoom.

>> **On/Off Labels:** To make it easier to see whether a setting is on or off, you can have Apple Watch show an additional label on the on/off switches.

>> **Grayscale:** To assist users for whom color might impair visibility, Apple Watch lets you enable grayscale onscreen. After you set the filter, the settings apply systemwide.

>> **Visual enhancements**: Apple Watch gives you several ways to enhance the visuals (including text) on your Apple Watch display. This includes Bold Text, Adjust Text Size, Reduce Motion, Reduce Transparency, Increase Brightness, and X-Large watch face. See Figure 2-14 for an example of making adjustments.

Hearing

For the past few years, hard-of-hearing and deaf Apple Watch users have been able to adjust the way they receive alerts and reply to messages. The following list shows the available options for hearing-impaired users:

>> **Mono Audio:** Users who are deaf or hard-of-hearing in one ear may sometimes be unable to hear certain signals when they're using Bluetooth headphones, because stereo recordings usually have distinct left- and right-channel audio tracks. Apple Watch can help by playing both audio channels in both ears, and letting you adjust the balance for greater volume in either ear. You can enable and adjust this setting.

FIGURE 2-14:
You can adjust screen brightness and text size in the Apple Watch itself. Tap the Settings icon, then tap Brightness & Text Size.

>> **Sounds & Haptics:** Apple Watch's Taptic Engine provides a gentle tap on your wrist every time a notification comes in. You can enable Haptic notifications or turn on Prominent Haptic to pre-announce some common alerts. To set this up, go to Settings ⇨ Sounds & Haptics on your Apple Watch. (See Figure 2-15).

>> **Scribble:** If you're not using Voice Dictation and you can't find the right Smart Reply to send as a message, you can write back by scribbling letters on the display. Apple Watch then converts those scribbles to text.

To send or reply to a message with Scribble, follow these steps:

1. *On your Apple Watch, open a text message or email, scroll to the bottom of the message, then tap the Scribble icon.*

2. *With one finger, write your reply, then tap Send in the upper-right corner of your screen.*

Physical and Motor Skills

As you likely know, Apple Watch features several fitness algorithms and sensors, but did you know this also includes support for wheelchair users, helping them more accurately track their activity?

For example, when the Wheelchair setting is enabled, pushes are detected instead of steps, and Apple Watch will register different types of pushes, speeds, and terrains. You'll also see a Roll goal in the Activity app instead of a Stand goal. (See Figure 2-16.)

FIGURE 2-15:
You can make adjustments on the Apple Watch app on iPhone or on the watch itself. *Taptics* refer to the slight buzz or tap you feel on your wrist. There are many ways to customize what you feel, when, and at what strength.

As I discuss in Chapter 8, there are wheelchair-specific workout options: Outdoor Wheelchair Walk Pace and Outdoor Wheelchair Run Pace. Just choose a workout and Apple Watch turns on the appropriate sensors.

To turn on the health and fitness features designed for wheelchair activity, edit the Wheelchair preference in the Health section of the Apple Watch app, as follows:

1. **Open the Apple Watch app and tap Health.**

2. **In the upper-right corner, tap Edit, then tap Wheelchair.**

3. **Select Yes, then tap Done in the upper-right corner.**

FIGURE 2-16:
Apple Watch's many accessibility features include an Activity app optimized for wheelchair users (left), which can remind you to move, too (right).

Some other accessibility options tied to dexterity and mobility:

>> **Quick Actions:** New in watchOS 9, users can double pinch (their forefinger and thumb together) to perform a quick action.

>> **Auto-Answer Calls**: When a call comes in, this tab lets you auto-answer them (regardless of the caller).

>> **Side Button Click Speed:** Selects the side button click speed (default, slow, and slowest).

>> **Touch Accommodations:** If you have trouble using the touchscreen, adjust some settings here to change how the screen responds to touches.

>> **AssistiveTouch:** To support Apple Watch users with upper-body limb differences, AssistiveTouch enables one-arm usage of Apple Watch. When activated, it senses hand gestures to interact without touching the display. Thanks to its built-in motion sensors, Apple Watch wearers can answer incoming calls, control an onscreen motion pointer, start a workout, initiate an action menu that can access Notification Center, Control Center, and so on.

Chapter **3**

Control Freak: Mastering Apple Watch's Interface and Apps

Are you ready for a deeper dive into Apple Watch's controls, features, and options? That's precisely what I cover in the following pages — or digital pages, if you're reading *Apple Watch For Dummies:* 2023 Edition, in ebook form.

Though it's unlikely that you'd want to read an electronic book on Apple Watch's diminutive screen, your always-on mobile companion can do so much to help you throughout the day. This chapter explores what's possible, beginning with mastering the controls.

As with many other Apple products, the user experience is paramount, and the Cupertino, California–based company has nailed the interface once again with Apple Watch.

In other words, Apple Watch is easy to use.

After you look at using Apple Watch's screen and buttons (and microphone and speaker), I suggest a few ways to take advantage of the watch's main features, including Dock, Control Center, and Notifications. Then I wrap up with an overview of the built-in Apple apps.

TIP

Although some of the Apple Watch models are best used while wirelessly tethered to a nearby iPhone, the Series 6, Series 7, and Series 8 watches do have 32 gigabytes (GB) of storage. (Fun fact: Storage was 8GB for the first couple of models and then 16GB until Series 6.) Apple uses the integrated storage for various things, including watchOS (operating system) and parts of installed apps. But you can store quite a bit of offline music and podcasts; therefore, you can leave your phone at home yet still <u>listen</u> if you go for a jog — although Bluetooth headphones, such as Apple's trendy AirPods, are recommended (or you can stream Apple Music if you own a GPS + Cellular watch model!). See Chapter 9 for more on music playback. You can also upload a lot of photos for viewing if your phone isn't around.

Handling Apple Watch's Controls

This watch is on your wrist, so the main way you interface with it is with your fingertips. But you have a few ways to do it.

Just like you do with your iPhone, iPod touch, and iPad, you can use your fingers on the Apple Watch screen to tap, double-tap, press, two-finger press, and swipe. The Digital Crown button and the side button also help you access myriad features on your Apple Watch.

Tap

Tapping something, such as an icon, is akin to pressing Enter on a computer or clicking the left button on a computer mouse. Tapping confirms a command, as in "Yes, I want that" — whether it's playing a song, accepting a calendar invitation, starting turn-based directions on a map, or hanging up on a completed call. As shown in Figure 3-1, tapping is the most frequent method for interacting with Apple Watch.

Speaking of taps, you can send a tap to friends and loved ones to let them know you're thinking about them. Apple calls this feature *Digital Touch*; it's a communications option that lets you send an image and subtle vibration to someone else's Apple Watch (required). See Chapter 5 for more on sending taps to friends and loved ones.

Double-tap

You might not double-tap very much, but Apple Watch does recognize this action. First, you must enable Accessibility settings by triple-tapping the Digital Crown button. (See Chapter 2 for more on these settings.). When you do so, you can activate *Voice Over*, a feature that speaks onscreen text to you in a humanlike voice, simply by double-tapping the screen. Another Accessibility option is Zoom: Quickly tap the screen twice with two fingers to get a closer look at text or images.

Press (Force Touch)

Apple Watch understands the difference between a light touch and a deep press and reacts accordingly. That is, you can tap quickly, as explained in the "Tap" section, or press and hold for a second to access a range of contextually specific controls tied to the task at hand. Apple calls this feature *Force Touch*. When you use Force Touch, pressing the screen firmly displays additional controls in such apps as Messages, Music, and Calendar. It also lets you select different watch faces, pause or end a workout, search an address in Maps, and more. Force Touch is the most significant new sensing capability since Multi-Touch on smartphones!

Two-finger press (time or heartbeat)

If you lightly press two fingers on the Apple Watch clock face screen, a voice announces the time for you.

Also, you can let someone special know you're thinking about them by sending that person your heartbeat, as shown in Figure 3-2. The catch? The recipient also needs an Apple Watch.

In the Messaging app (see Chapter 5), tap the Heart icon and then press two fingers on the screen at the same time. The built-in heart-rate sensor records and sends your heartbeat to a loved one, who will feel it on their wrist. Apple calls this feature "a simple and intimate way to tell someone how you feel," and it's not available on any other smartwatch.

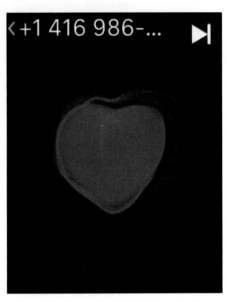

FIGURE 3-2:
Send your heartbeat by pressing two fingers on the screen at the same time.

Flip to Chapter 5 for more details on sending your heartbeat to someone.

Swipe

Like most other mobile devices in your life — such as a smartphone, tablet, and many laptops — Apple Watch supports onscreen swiping. If the app supports this feature, you can swipe around with your fingertip. If you own an Apple Watch Series 7 or Series 8, see the section below, called Type, for swiping across the QWERTY keyboard.

Sketch

Just as you can tap, press, and swipe on the Apple Watch screen, you can use your finger to draw something on the small display and then send it to someone else who wears an Apple Watch. Apple calls this type of drawing a *sketch*. A sketch can be a flower, heart, puppy, butterfly, wedding ring, Christmas tree, or whatever; see Figure 3-3 for an example.

The person to whom you send your sketch can not only see what you drew, but also watch your sketch come to life with animation — and then respond with a sketch of their own. See Chapter 5 for more on sketches.

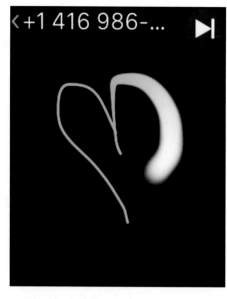

FIGURE 3-3:
Sketch whatever you like — in your desired color — and then send it to another person's Apple Watch.

Type

Those who own an Apple Watch Series 7, Apple Watch Series 8 or Apple Watch Ultra can pull up a virtual (onscreen) QWERTY keyboard — yep, just like the layout your laptop's keyboard has — if this layout is more familiar and comfortable for you. Why is this option available in these newer models and not previous ones? It's because the screens on the latest Apple Watch models are larger than their predecessors.

To use this feature, simply tap the open "reply" area of a message or email, which pulls up the QWERTY keyboard. Then you can tap each letter you want to type.

Another way to use this virtual keyboard is to swipe your finger between letters to form your words, a feature that Apple calls *QuickPath*. Artificial intelligence attempts to guess what words you're looking to type, which speeds your message-writing considerably.

REMEMBER

Apple has added several "accessibility" features to Apple Watch over the years, as you see in Chapter 2. But because this is a chapter about interacting with your Apple Watch, here's a friendly reminder that AssistiveTouch enables one-arm usage of Apple Watch, which senses hand gestures to interact without touching the display.

Digital Crown button

Apple is known for introducing new ways to interact with content, and the Digital Crown button is no exception. If the watch is worn on the left wrist, this button is on the top-right side of the Apple Watch case. This button looks like the crown on mechanical watches, which is traditionally used to wind the main spring and to set the time. For the watchOS platform, the button is used primarily to magnify content on the small screen without your fingers getting in the way of content. Instead of pinching to zoom, as you would on an iPhone or iPad, you twist the Digital Crown forward or backward to zoom in and out of photos or maps. You can also use it to scroll quickly through contacts, songs, and other lists.

Keep in mind that you can flip the band around to wear the watch on your right wrist, which places the Digital Crown on the left side of Apple Watch. You can keep the side button and the Digital Crown button on the right side of the watch while wearing the watch on your right wrist, of course, but it may not be comfortable for you to access these buttons with your left hand. Go to the Settings area to change which wrist you're wearing your watch on. (See Chapter 11 for more on changing the watch's orientation.)

At any time, you can press the Digital Crown button to see the watch face or (if the watch face is already visible) return to the Home screen, which is similar to pressing the Home button on an iPhone or iPad. (If you have one of the newer iPhones and iPads, which don't have a physical Home button, pressing the Digital Crown button is like swiping up from the bottom of the device to get to your Home screen.) Or press and hold the button to activate Siri. Double-pressing the Digital Crown button switches between the watch face (the one you've chosen to display) and the last app you used. This feature is a fast and convenient way to see the time, regardless of the app you're in. Twist the Digital Crown button to zoom, scroll, or adjust what's on the screen.

To summarize, here's how to use the Digital Crown button for specific tasks:

>> **To return to the Home screen:** Press the Digital Crown button to return to the watch's Home screen (similar to pressing the Home button or swiping up from the bottom of the screen on an iPhone or iPad).

>> **To return to the clock or return to the most recent app:** Press the Digital Crown button to see the clock again. Double-press the Digital Crown button to go back to the last app you were in.

>> **To activate Siri:** Press and hold the Digital Crown button to launch your voice-activated personal assistant. Or just say "Hey, Siri" into your watch and follow with a question or command.

>> **To zoom/scroll:** On the Home screen and in supported apps, twist the Digital Crown button forward or backward, as though you were winding a mechanical watch, to scroll through lists or zoom in on a photo, a map, the Home screen, and more.

>> **To unlock the screen:** If you're a swimmer, you can turn the button to unlock the screen after a swimming workout (Apple Watch Series 2 or later).

Side/power button

Apple Watch has another handy button that's located just below the Digital Crown button on the right side (if you're wearing Apple Watch on your left wrist). Depending on whom you ask, this button is called the *side button* or the *power button*.

REMEMBER

You can wear Apple Watch on your right wrist and turn the watch case around, which puts the side button (and Digital Crown) on the left side of the watch. See Chapter 11 for more on changing the watch's orientation.

Some people refer to this side button as a power button because holding it down for a couple of seconds displays a power-down screen, where you can choose to turn your watch off. This process is similar to pressing and holding the power button on the top or side of an iPhone or iPad.

Double-pressing the side button initiates Apple Pay, which allows you to use your Apple Watch to make a purchase at a participating retailer or from a compatible vending machine. After you pair your Apple Pay account with a credit or debit card, NFC (near field communication) enables you to wave your smartwatch over a contactless sensor to initiate the transaction. You should hear a faint tone and feel a slight vibration to confirm that this digital handshake is complete. Apple Pay works on Apple Watch as long as the watch stays in contact with your skin. See Chapter 10 for more on Apple Pay.

Double-tapping the side button also lets you select a card that's synced with the Wallet app on your iPhone. You can sync boarding passes, event tickets, coupons, student ID cards, proof-of-vaccination cards, and much more. See Chapter 10.

Finally, pressing and holding the button initiates the SOS feature, which allows you to call for help quickly and easily and alert your emergency contacts that you're in danger.

Following is a summary of the side button's features:

>> **Power:** Press and hold the button until you're prompted to turn off the power. You probably won't use this feature very much, just as you don't power down your iPhone or iPad often, but you can if you want.

>> **Pay:** Double-tap the side button to launch Apple Pay when you're about to buy something. You can also access cards synced with the Wallet app on iPhone (such as boarding passes or coffee shop loyalty cards).

Apple Pay requires skin contact to operate, so ensure that the watch is snug on your wrist before you press the side button twice to launch Apple Pay.

REMEMBER

>> **SOS:** Press and hold the side button to display and activate SOS, if you need to have your watch call an emergency contact or emergency services. If you don't have a cellular Apple Watch option, you require a nearby iPhone for these emergency numbers to be dialed. Chapter 8 has more on the SOS feature.

Action button (Apple Watch Ultra only)

Available in Apple Watch Ultra — Apple's largest and most durable Apple Watch, to date — is a dedicated orange Action button on the left side of the watch that can be used to quickly access various apps or features within apps. You can customize what this button does (such as marking your location on a run), but by default it emits an 86-decibel emergency siren to alert people in your vicinity.

Going Hands-Free with Siri

Can you think of an even more natural way to interface with a smartwatch than touching it?

How about talking into it?

Apple's Siri (pronounced "sear-eee") is a voice-activated personal assistant that lets you ask questions or give a command on an iPhone, iPad, or iPod touch — and now Apple Watch. But unlike the other iOS devices, Apple Watch allows you to simply press and hold the Digital Crown button to ask a question (and you have other hands-free ways to activate Siri, as you'll see shortly).

In Chapter 7, I cover all the ways that Siri can help you master Apple Watch, but for now, I offer a snapshot of how to use it.

To activate Siri on your Apple Watch, follow these steps:

1. **Press and hold the Digital Crown button.**

   You see something like **What can I help you with?** on your Apple Watch screen. Siri is ready for your instructions after the short chime.

2. **Ask Siri a question or give it a command.**

   You could ask "Who's winning the New York Yankees game?" In fact, you don't have to say the full team's name; therefore, asking how the "Yankees" are doing rather than the "New York Yankees" is usually fine. Siri shows you the requested information or tells you what you asked for.

   As the information is being displayed — in real-time, no less — Siri also says something like "Okay, sports fans, let's have a look" or "Here you go."

3. **Press the Digital Crown button again if you want to go back to the Home screen.**

   You should see Apple Watch's Home screen, flush with icons.

REMEMBER

As you do on an iPhone, iPad, or iPod touch, you need an Internet connection to use Siri on Apple Watch. The requested information — like your question about the baseball score — is sent to Apple's servers for processing, and the answer is sent to your watch. Unless you have the cellular model and have set up your monthly plan, your watch needs to be connected to a nearby iPhone via Bluetooth or Wi-Fi.

Vibrating Along with Apple Watch's Tactile Feedback

Apple Watch can be tapped, pressed, swiped, and spoken to. But your smartwatch can also tap *you* in the form of a light vibration. Powering this technology is what Apple calls a *Taptic Engine.*

Apple's Taptic Engine is a linear actuator inside Apple Watch that produces haptic feedback. In plain English, this technology produces a slight buzz on your wrist whenever you receive an alert or notification or when you press down on the display (a feature that some smartphones offer). Another comparison may be a video game controller that buzzes in your hands in conjunction with what's happening on your TV screen (such as your character's being shot in a first-person shooter).

The Taptic Engine not only gives you information without your having to look at your wrist — such as when to turn left or right (based on the number of taps you felt) while you use Apple Maps — but also enables new and intimate ways to interact with those who also own an Apple Watch. Chapter 5 further discusses how to send a tap or your unique heartbeat to a friend or loved one.

Using Control Center, Dock, and Notifications

Apple Watch gives you a few ways to glean information on the go: Control Center, Dock, and notifications.

Keep in mind that Apple Watch isn't meant for reading lengthy websites — the watch doesn't have a web browser — because it's designed for quick interactions. Though they differ, Control Center, Dock, and notifications give you bits of customized information when and where you need them.

Control Center

In an earlier version of watchOS, swiping up from the bottom of the screen launched Glances, which were snippets of information, such as how your favorite sports team was doing, what the weather was like, or how your stock was performing. Now this motion initiates Control Center, which displays information about your watch, such as battery percentage, Wi-Fi and cellular signal, flashlight mode (a feature that turns the screen bright so you can see in the dark), Airplane mode, and more than a half-dozen other options (see Figure 3-4).

To activate Control Center on your Apple Watch, follow these steps:

1. **Swipe up from the bottom of the clock (time) screen.**

   You'll see icons for various actions and pieces of information, including

   - Remaining battery percentage

   - Whether your Apple Watch is using Wi-Fi or cellular

   - Whether your flashlight app is on

   - Whether Water Lock (which pushes water out of the speaker after you've been swimming) is enabled

   - A feature that pings your nearby iPhone

- Whether Airplane mode is on

- Enabling Walkie-Talkie mode (see Chapter 5)

- Whether Theatre mode is activated (keeping the watch silent and the screen dark until you tap it)

- Where to play audio (Apple Watch or a wirelessly connected device, such as AirPods)

2. **Select one of the Control Center options.**

   Tap one of the dozen or so options to turn it on, and tap again to turn it off. Can't find your iPhone? Tap the icon that looks like a phone with sound coming out of it; your Apple Watch pings your missing iPhone to help you locate it nearby (with Bluetooth, generally 30 to 50 feet).

3. **Tap Edit to customize Control Center.**

   You can change what you see on your Dock, for example, by scrolling to the bottom of Control Center and selecting Edit. The icons wiggle, which means that you can move them "drag and drop" style into a new order on the Apple Watch screen.

Apple Watch

Apple Watch with Cellular

FIGURE 3-4: Access Control Center on Apple Watch by swiping up from the bottom of the screen. Keep swiping to see different icons that you can use to make changes quickly.

Dock

Over the past couple of years, Apple added a handy Dock feature for Apple Watch wearers. You activate this feature by pressing the side button. Quite simply, Dock lets you open your favorite apps or move from one app to another quickly.

Here's how to get going:

1. **Choose the apps you want to appear in Dock.**

You can select up to ten of your favorites, in fact. To choose what you want, follow these steps:

(a) *Open the Apple Watch app on an iPhone.*

(b) *Tap My Watch and then choose Dock.*

(c) *Tap Edit and then add or remove apps to choose your favorites.*

To rearrange apps, touch and hold next to an app, and then drag up or down.

(d) *Save your changes by tapping Done.*

2. **Press the side button on your Apple Watch to activate Dock.**

3. **Swipe up or down.**

Alternatively, you can turn the Digital Crown button to cycle through the apps you opened most recently or your favorite apps, as shown in Figure 3-5.

4. **Tap to open an app.**

If you scroll all the way down to the bottom of the screen, you can tap All Apps to go to the Home screen.

5. **Close Dock by pressing the side button again.**

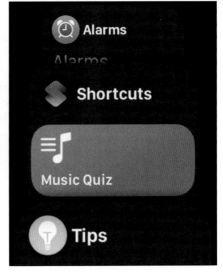

FIGURE 3-5: Press Apple Watch's side button to launch Dock, display your recently opened apps, or set Dock to display up to ten of your favorite apps.

Notifications

Much like notifications on an iPhone or iPad, Apple Watch notifications display app information at a specific time. A notification could be a voicemail waiting for you, a calendar appointment, a news headline from CNN, a social media notification (someone started following you or liked your photo), a comment made to your online classifieds ad, or a game letting you know that it wants you to come back.

Apple Watch has two kinds of notifications:

>> **Short-look:** These notifications appear when a local or remote alert arrives (see Figure 3-6) and present a minimal amount of information. When you lower your wrist, the short-look note disappears. Information includes the app name, icon, and title.

>> **Long-look:** These notifications appear when your wrist is raised or when you tap the short-look interface. You receive more detailed information and more functionality, such as four action buttons for additional information. You need to dismiss these notifications when you're done by swiping up.

Apple asks its app developers to make sure that notifications are relevant to what users want so that they're not bombarded with messages all day long. Nonetheless, you can turn off notifications for any app in the Apple Watch app on an iPhone; see Chapter 6.

FIGURE 3-6:
Notifications on your Apple Watch can show you a bit of time- or location-relevant information.

Looking at Apple Watch's Built-In Apps

Similar to the preinstalled apps designed to help get you started on other Apple products, Apple Watch has several built-in apps, which do the following things:

>> **To keep you in touch:** Phone, Messages, and Email

>> **To prevent you from getting lost and to find stuff around you:** Maps

>> **To stay organized and informed:** Calendar, Clock, Weather, and Stocks

>> **To entertain you:** Music and Photos

>> **To keep you healthy:** Activity, Workout, Mindfulness, Breathe, Sleep, and Cycle Tracking

I list some of these apps briefly in Chapter 1, but here, I discuss them in greater depth because you're likely to rely on many of them. Okay, some apps may not appeal to everyone, such as Stopwatch and World Clock, but they're available if and when you want them.

Phone

You no longer need to reach for your iPhone to see who's calling you. Simply glance at your wrist to see the name or number (if the caller isn't in your Contacts) and decide whether to take the call. You can have a chat through the watch, if you like, or transfer the call to your iPhone. Don't want to take the call? Simply cover the Apple Watch screen with your hand to mute it. You can also place a call to someone through Apple Watch, even if your iPhone isn't nearby, provided that you have a cellular plan for your supported Apple Watch or make the calls over Wi-Fi. See Chapter 5 for more on talking on your Apple Watch.

Messages

If someone sends you a text message (Short Message Service [SMS]) or an iMessage to your iPhone, Apple Watch gives you a subtle tap to let you know about it. Raise your wrist to see who wrote it and to read the message. You can reply with a preset response, send an animated emoji, dictate the response, type a reply on the virtual QWERTY keyboard (Apple Watch Series 7 only), or record and send a short audio message. Beginning with watchOS 8 is the option to edit the presented text in dictated messages. For added expression, you can enter a word or phrase in a message and choose among hundreds of trending animated GIFs. See Chapter 5 for more on handling messages on your Apple Watch. Figure 3-7 shows an example of a text message.

FIGURE 3-7:
If you're able to speak aloud, it's super-fast to use your voice to dictate a reply.

Mail

Although a small smartwatch screen may be more conducive for short text messages than lengthy emails, you can read your personal or professional email on your watch, as shown in Figure 3-8. You can also flag emails, mark them as read or unread, reply to them, or delete them; press and hold the screen to access a few options. But you can't create an email from scratch; to do that, you need your iPhone. See Chapter 5 for more on reading and responding to emails on your Apple Watch.

Memoji

Make your own Memoji (a term that fuses the words *me* and *emoji*), which is an animated cartoon representation of yourself you can use in messages, as shown in Figure 3-9.

Note: You can create Memoji on the Apple Watch with the Memoji app, or you can create them with the Messages app on an iPhone or iPad and then sync your devices.

Tap Get Started on Apple Watch, and start selecting skin, hairstyle, brows, eyes, head, nose, mouse, ears, facial hair, eyewear, and headwear. It's fun!

FIGURE 3-8:
Read your
email on Apple
Watch.

FIGURE 3-9:
Create a
personalized
Memoji.

Calendar

Your wrist can tell you when you've got an upcoming calendar appointment. You can set meeting reminders, accept or decline calendar invitations, and (if desired) email the organizer with a preset response. See Chapter 6 for more on Calendar options.

Alarm

Apple Watch lets you set and manage multiple alarms, such as those for meetings or wakeup calls. You can ask Siri to set them or set them yourself, using the Digital Crown button to tweak the alarm time. You can even set the alarm to vibrate on your wrist. Figure 3-10 shows the Apple Watch's Alarm app. Don't forget that you can also sync iPhone alarms with your Apple Watch. See Chapter 4 for more on setting alarms.

FIGURE 3-10:
Use the Alarm app to have your watch wake you up.

Stopwatch

Apple Watch offers a stopwatch for your convenience, much like the one on your iPhone; you can set it to digital, analog, or hybrid view. Apple Watch also offers an optional graph view. See Chapter 4 for more on using the Stopwatch app on Apple Watch.

Timer

Whether you're running around a track or cooking something in the oven, Apple Watch has a timer for that. As the timer runs, a line moves around the dial to give you a sense of how much time has passed and how much is left. See Chapter 4 for more on setting timers on your Apple Watch, including the option to set multiple timers, introduced in watchOS 8.

World Clock

You don't need to go to a website to see the time in cities around the world (or count on your fingers as you do the manual calculation between, say, New York and London). Launch the World Clock app to see the time in cities of your choice.

You can use your iPhone to add new locations at any time. See Chapter 4 for more on setting up the World Clock app.

Siri

Siri is a fast, convenient way to interface with your Apple Watch. In fact, nothing is more natural than using your voice; your smartwatch responds after you say, "Hey Siri," followed by a question or command. Alternatively, you can push and hold the Digital Crown button to activate Siri. See Chapter 7 for more on using Siri to do things on your Apple Watch.

Weather

Use Apple Watch to check the weather where you are or in any city in the world. This app shows the temperature and precipitation at that exact moment or in the days or week ahead. Beginning with watchOS 8, the Weather app now supports Severe Weather notifications (so it will display government alerts about major weather events), and it delivers Next Hour precipitation alerts and updated complications for watch faces. See Figure 3-11 for a glimpse at the Weather app. If you haven't chosen a watch face that includes weather information, the quickest way to access that information is to ask Siri for it. The second-quickest way is to swipe up and access the Weather Glance screen. Opening the app itself, of course, is the third-quickest way, but you should choose the method that works best for you. See Chapter 6 for more on the Weather app on your Apple Watch.

Stocks

Follow all the companies in which you have a vested interest — or plan to have one — with the Stocks app on your Apple Watch. This app lets you keep up with the stock market price, point and percentage, market cap, and more. All your stocks include a historical graph. See Chapter 6 for more on the Stocks app on your Apple Watch.

Activity

Activity is one of the two important fitness-related Apple Watch apps. The three Activity rings — Move, Exercise, and Stand — give you a simple yet informative glimpse of your daily activity goals and progress. The app is designed to motivate you to sit less and move more. See Figure 3-12 for the three Activity rings and Chapter 8 for more on the Activity app.

FIGURE 3-11:
We all love knowing about the weather, and Apple Watch provides multiple ways to see it, including the Weather app.

FIGURE 3-12:
The smart, color-coded Activity app shows your daily progress.

Workout

During an exercise routine, the Workout app shows you real-time fitness information, including time, distance, calories, pace, and speed. You can choose preset workouts, such as Running, Walking, Cycling, and others. Figure 3-13 shows you the newest additions to the Workout app (in watchOS 9): new ways to view your info, heart rate zones, new and more customized workouts, "ghost" runs (to compete against your last or best results), and more. Chapter 8 provides more information on using Workout on your Apple Watch.

FIGURE 3-13: Updated for watchOS 9, the Workout app now supports many ways to view your workouts, by simply twisting the Digital Crown.

Maps

Whether you're trying to find a restaurant in your hometown or feel like going on a stroll in an exotic city, your Apple Watch can give you directions based on your current location, as shown in Figure 3-14. See the fastest route, get turn-by-turn navigation instructions (including taps on your wrist when it's time to turn), or ask Siri to find local businesses. See Chapter 6 for more on the Maps app on your Apple Watch.

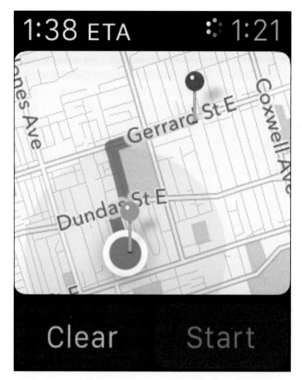

FIGURE 3-14:
Get directions
to a location,
such as a
business,
by using the
Maps app.

Photos

Now those who matter to you are just a glance away. The Photos app on Apple Watch displays your photos of loved ones, friends, pets, scenery, and other memories. Use the Digital Crown button to zoom in on individual images or swipe to browse through your collection one photo at a time. You can even choose photos to load on your watch, even if your iPhone isn't nearby. See Chapter 12 for more on using the Photos app on your Apple Watch.

Camera Remote

Although Apple Watch doesn't have a camera, you can use it as a live viewfinder for your iPhone's rear-facing camera. Therefore, you can use your watch to see and snap a subject, essentially turning your wrist into a wireless shutter (ideal for selfies). Also, you can adjust the timer remotely. See Chapter 12 for more on using your Apple Watch to control your iPhone's camera.

Music

You can keep your iPhone tucked away in your pocket or purse and play your music remotely on your Apple Watch. Figure 3-15 shows what your songs might look like on your watch, complete with album artwork if you have it on your iPhone. You may want to load your watch with a few hundred tunes to listen to (Bluetooth headphones required) when you don't have your phone with you. Beginning with watchOS 8, the Music app lets you share songs, albums, and playlists through Messages and Mail, and enjoy music and radio all in one place (there used to be a dedicated Radio app for Apple Watch). See Chapter 9 for more on listening to and managing your music on your Apple Watch.

FIGURE 3-15: Apple Watch can store music you synced from your iPhone. You can also control tunes on your iPhone or stream directly from Apple Music.

Remote

You can not only control your iPhone music remotely via your Apple Watch, but also use it to access a nearby Apple TV box connected to your TV. Your wrist-based remote can navigate the main menu, scroll through media lists, and select what you want. You can also use the watch's Remote app to control your iTunes library and iTunes Radio on a PC or Mac. See Chapter 9 for more on using your Apple Watch as a remote control.

Wallet and Apple Pay

Formerly referred to as Passbook, this handy app for iPhone also works well on Apple Watch; it lets you keep track of such things as boarding passes, movie and theater tickets, and loyalty cards. With watchOS 8, you can even unlock your car from a distance and start it from the driver's seat (for compatible vehicles and Apple Watch Series 6 or newer). In addition, you can add keys for your home, office, and hotel to Wallet, and tap the Apple Watch to unlock the doors. When you shop with Apple Pay, you choose which credit or debit card to use for the purchase. See Chapter 10 for more on using Apple Pay and Wallet with your Apple Watch.

Mindfulness

Live a better day by taking a minute to "Breathe" and "Reflect." Figure 3-16 is what you'll see when you tap the Mindfulness app on Apple Watch. Formerly Breathe, the new app nudges you to take a small mental health break during the day. Breathe is still here, with an enhanced experience, along with a new session type called Reflect, which welcomes you with a thoughtful question or suggestion to put you in a positive frame of mind. For example, it might be "Think about a relationship you cherish and why it matters to you so much." You can twist the Digital Crown button to change the frequency and duration of the breathing and reflecting exercises the app guides you through.

TIP

watchOS 8 (and later) supports Focus (previously available only on the iPhone), which Apple says "helps you reduce distractions and be in the moment." Your watch automatically mirrors any Focus settings on your iPhone, so notifications from people and apps are filtered based on what you're currently doing.

FIGURE 3-16:
Take a moment out of your hectic life to focus on your breathing with the Mindfulness app, which folds in two experiences: Breathe and Reflect.

Sleep

Use this app to schedule your wake-up alarm, bedtime, and bedtime reminders on your Apple Watch or iPhone. You can also set up a sleep goal, such as trying to get the recommended eight hours per night (I wish!); your Apple Watch tells you whether you've hit your goal. New with watchOS 8 is the ability to track your sleeping respiratory rate (how many breaths you take per minute). Apple Watch uses its built-in accelerometer to measure your respiratory rate while you're sleeping, and you can view this information, along with trends over time, in the Health app on your iPhone. So cool. New with watchOS9, sleep stages: See how much time you spent in REM, core, or deep sleep, as well as when you might have woken up.

Walkie-Talkie

Communicate Apple Watch to Apple Watch. This app lets you have a voice conversation with someone else via your watches. Any model of Apple Watch works — even Series 1 — but you need Apple watchOS 5 or newer to use it. Contacts you add, such as your partner, kids, or boss, can talk to you whenever you're set to Available status. You'll see some suggested contacts (such as family members) and then a list of all the people listed in your iPhone's Contacts app. Tap a contact to invite them, and they'll have a chance to accept. Swipe down to show whether you're available to chat. I take a deeper dive into Walkie-Talkie in Chapter 5.

Heart Rate

As you might expect, tapping this little red heart shows your current heart rate and daily summaries on Apple Watch. This Health app can notify you if Apple Watch detects that your heart rate is high or low. You can adjust or turn off these notifications in the Apple Watch app on your iPhone. Tap Turn On to activate this feature, and you'll see it measuring your heart's beats per minute (BPM) and logging the time your rate was measured, as well as some historical info.

Medication

Part of the Apple Health app, you'll find a new icon on your Apple Watch for Medications, which lets you log (and view) your meds right from your wrist. By having access to this info at a glance, you can conveniently keep track of your meds (as well as vitamins and supplements) and receive reminders on when to take them.

ECG

As you'll read about in Chapter 8, the ECG (electrocardiogram) app can record your heartbeat and rhythm using the electrical heart sensor on Apple Watch Series 4, Series 5, Series 6, or Series 7 — and then check for atrial fibrillation (AFib), a form of irregular rhythm.

Blood Oxygen

As you see in Chapter 8, this app starts a 15-second blood-oxygen measurement (see Figure 3-17). Small beams of light shine from the back of your watch into your body, measuring the amount of oxygen in your blood cells. Cool! Generally speaking, a healthy reading is between 95 and 100 percent. This feature is available only in Apple Watch Series 6 and Series 7.

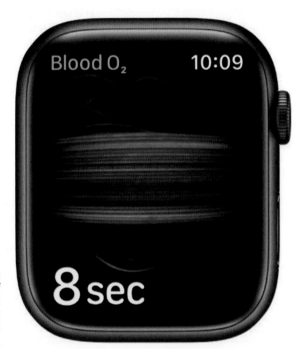

FIGURE 3-17:
Keep track of your oxygen level with the Blood Oxygen app.

Find People

The little green-and-blue icon launches the Find People app. The first time you tap this icon, the app asks your permission to access your location on a map while it's in use and to allow people whom you select to see you. Then the app shows your location on a map, and you (and they) can see how far away friends are. Tap Allow or Don't Allow when you're prompted (first time only).

Find Devices

Just as your Apple Watch can help you find people or items (via AirTags), it can also be used to locate other Apple devices, such as an iPhone, iPad, iPod touch, Apple Watch, AirPods, or Macs. Introduced in 2021, the Find Devices app may help find an Apple product if misplaced — so long as they're connected to the same Apple ID. Chapter 6 shows you how to set up and use this handy app.

Podcasts

In case you haven't heard, podcasts are free downloadable programs, which can be comedy routines, news reports, political rants, tech advice, religious sermons, or the latest music remixes from a popular club DJ. Podcasts can have video too, but only audio versions work on Apple Watch. When you subscribe to a podcast, the show automatically downloads to your device so that you can hear it on-demand. The Podcasts app on Apple Watch syncs with your iTunes library (on PC) or Apple Music app (for Mac) so you can hear everything from your wrist (presumably via wireless headphones). See Figure 3-18.

FIGURE 3-18:
Sync or stream your favorite podcasts right to your wrist — and then to your Bluetooth headphones.

Audiobooks

The Audiobooks app, shown in Figure 3-19, lists and plays all the books on your iPhone (inside the Apple Books app). Neat, huh? That is, all titles in your Reading Now list are synchronized to your watch. Now you can listen to a chapter while you're out for a jog and continue on your iPhone or iPad when you get home, as the app remembers where you left off. After you open the Audiobooks app, you have options on how to play acquired audiobooks.

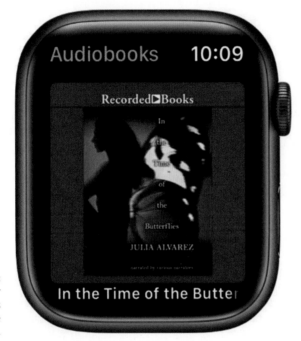

FIGURE 3-19:
Play all your audiobooks on your Apple Watch.

App Store

You no longer need an iPhone to browse and download Apple Watch apps to your wrist! Simply tap the large *A* icon on your Apple Watch home screen to launch the Apple Watch App Store, shown in Figure 3-20. Then start browsing (checking out trending or featured apps, for example) or searching (using Scribble or Dictation), along with looking at screen shots, reading reviews, and checking prices. It's about time for us to access the App Store directly on Apple Watch!

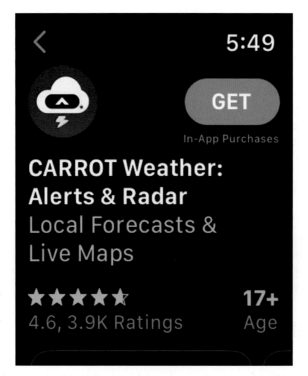

FIGURE 3-20:
Browse or search and download new Apple Watch apps directly to your device.

Noise

High noise could damage your hearing. The Noise app on Apple Watch has an ear out for you, as it can sense when ambient noise levels reach a dangerous level. When you're at a rock concert or working with heavy machinery, the app might suggest that you put in some earbuds; tapping your wrist shows you the estimated decibel level and duration of your exposure. Chapter 8 has more information on the Noise app.

Cycle Tracking

Cycle Tracking (which I cover more in Chapter 8) provides insight into your menstrual cycle and helps paint a clearer picture of your overall health. Along with data that you may want to share with a physician — flow information (Light, Medium, or Heavy), symptoms (such as cramps or headaches), and cycle length — you can see a graphical chart on your iPhone's Health app. The app can even alert you when it predicts that your next period or fertile window is about to start.

Home

Smart homes all around the world are increasingly powered by Apple HomeKit–enabled products, which means that these supported devices can talk to one another securely. You use an app or your voice (through Siri) to control everything. For example, you could say "Hey Siri, turn on the lights" or "Lower the thermostat by two degrees." Apple Watch lets you control your smart home on your wrist, too. Tap the orange-and-white Home icon, and all your HomeKit items are listed.

Just remember that you need to add the Home app on an iPhone first to get things to work.

Redesigned in 2021, the Home app, shown in Figure 3-21, lets you view who is at the door when it's synced with an Apple HomeKit–enabled camera. What's more, you can tap Intercom to broadcast a message throughout the home via smart speakers, such as HomePod and HomePod mini.

Calculator

Now you can use a dedicated app on your wrist to access a calculator (a quick way to calculate a tip, for example), add up some receipts or checks you're depositing into a bank machine, and more. The calculator has the familiar orange, black, and white virtual buttons as the iPhone version. Don't forget that you can raise your wrist to activate Siri and ask a math question as well!

Voice Memos

Did you come up with a great idea that you want to capture? Or do you need to record a meeting for work or a lecture for school? Whatever your reason for wanting to record audio, you can do it on your Apple Watch, thanks to the dedicated Voice Memos app. Simply tap to launch, press the large red Record button, and tap again to stop. All recorded memos will be synced to your iPhone, iPad, and Mac.

Did you know? Although it's not an app per se, crash detection is available by default on Apple Watch Series 8, Apple Watch SE, and Apple Watch Ultra (as well as the iPhone 14 family)! If you're met with a sudden accident or crash when driving or riding in a passenger car, SUV, or pickup truck, both the iPhone and the Apple Watch give you the option to activate Emergency SOS, provide your location, and notify the emergency contacts. If you don't respond to this prompt 10 seconds after the impact, the device will automatically turn on Emergency SOS. The Crash Detection feature also works if you turn on Low Power mode on Apple Watch or iPhone.

FIGURE 3-21: The Home apps lets you access your compatible smart home gear, such as seeing who is at your front door or broadcasting a message to the family via smart speakers.

Settings

The Settings app for Apple Watch lets you enable or disable several settings, including Airplane mode (turning off all wireless radios), Bluetooth, and Do Not Disturb. This app also mutes your watch in case you don't want to hear any sounds from it. Did you lose your iPhone under the cushions? Your Apple Watch can make your iPhone ping loudly so that you can hear and find it. See Chapter 11 for more on the Settings app on your Apple Watch.

Compass

The built-in magnetometer inside the watch detects magnetic north and then automatically adjusts to show true north, so you now know which way you're facing in the Maps app (which can be quite handy). The stand-alone Compass app also shows additional information — heading, incline, latitude, longitude, and current elevation — or you can have compass information as a watch-face complication (see Chapter 4).

WARNING

Apple says that bands with magnetic clasps (as opposed to leather or silicone bands) may cause some interference with the built-in compass, which could limit its accuracy.

Shortcut

As the name suggests, a *shortcut* is a quick way to run actions, such as "It's TV time," which dims your lights and turns the television and sound bar on. You might create a shortcut called "Heading to Work" to get your estimated time of arrival, hear about your first calendar event, and start an Apple Music playlist.

Apple News

It's news you can use. The large *N* icon, in red and white, brings you curated headlines and stories from several publications (see Figure 3-22) that you can tweak. You can start on your Apple Watch and continue on your iPhone or iPad, or vice versa.

After all, you might find yourself in line at a supermarket or waiting for a friend on a park bench; tapping this red icon lets you view some of the top headlines while you wait. If something interests you, but you don't have time to read it now, tap Save for Later; the entire article will appear in the Apple News app on your iPhone. News sources include *The Washington Post,* Fox News, CNN, *Newsweek, Billboard, VentureBeat,* and many others.

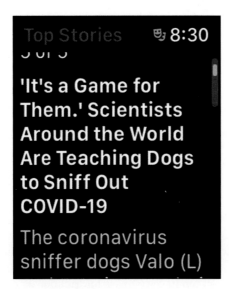

ɔ ʊi ɔ

'It's a Game for Them.' Scientists Around the World Are Teaching Dogs to Sniff Out COVID-19

The coronavirus sniffer dogs Valo (L)

Top Stories ⚇ 8:35

TIME

5 of 5

'It's a Game for Them.' Scientists Around the World Are Teaching Dogs

FIGURE 3-22:
Read the news wherever you are.

Contacts

The Contacts app is now on Apple Watch, too, mirroring your iPhone, and serving as a fast, easy way for you to browse, add, and edit contacts. You can even share contacts directly from the app.

2

Just the Tasks, Ma'am!

Explore the many Apple Watch faces, as well as how to customize them and add complications to them. You also discover how to turn your Apple Watch into an alarm, stopwatch, and timer.

Use your Apple Watch to make and take phone calls, send and receive text messages, and read and manage your email. You also discover how to send your actual heartbeat to a friend or loved one and use your watch as a walkie-talkie.

Turn your Apple Watch into a miniature media outlet by adding notifications on topics you care about, such as the current weather, stock market performance, breaking news, live sports scores, and more. Use the Dock to quickly access your favorite apps. Master shortcuts to get more done in less time. Then see how to keep track of appointments with the Calendar app and how to use your Apple Watch as a GPS via the Maps app.

Chapter **4**

It's About Time: Setting Watch Faces, Alarms, Timers, and More

A pple Watch is a watch, after all, so chances are that you'll use it a lot to tell time. But unlike a traditional analog watch or even most digital watches, Apple Watch lets you choose the face you want. This way, you can select what you'd like to see and how you'd like the information to be displayed.

This chapter looks at the multiple watch faces you can choose from, as well as customize so that you can truly make Apple Watch your own. You also see different ways to access the watch face on Apple Watch. Then you see other time-related apps preinstalled by Apple, including the World Clock, Stopwatch, Timer, and Alarm apps.

Looking At the Built-In Watch Faces

Why be stuck with only one watch face when you can have several?

That's one reason why people like a smartwatch. Because you've got a screen that can show virtually anything, you can go with a classic analog face (yes, with moving hands!), a digital watch face (just numbers), an analog–digital hybrid, or even one with animation. Some faces give you a ton of information so you can get caught up with a single glance at your wrist.

It's a breeze to change these watch faces whenever you like (see "Choosing Among the Various Watch Faces" later in this chapter) and without having to open the Apple Watch app on your iPhone. By default, your watch face is the one called *Metropolitan* (for Apple Watch Series 8). With this default setting, you get other onscreen information, called *complications,* which you can add or remove, including date, activity, and music.

TECHNICAL STUFF

You might be swept away by the many watch faces you can choose from and customize. To see the latest set of watch faces, make sure that your software is up to date. Look for software updates on your Apple Watch, and remember that the set of watch faces might differ from what you see on your Apple Watch, as not all watch faces are available in all regions, and some work only with newer versions of the watch (See the Apple Watch User Guide at https://help.apple.com/watch for more information.)

You can download many more watch faces from the Apple Watch App Store. These faces, in the form of apps, make Apple Watch uniquely yours, such as the virtual Cuckoo Clock app. To get you going, Apple has installed a few good (and customizable) built-in watch faces.

TIP

With watchOS 9 — the operating-system update that debuted in the fall of 2022 — you now have new ways to mark time. The Taptic Engine (that slight vibration on your wrist) can silently tap the hour on your wrist, if you like, so that you'll know the time even without looking at your watch. Or you can set a chime to ring every hour on the hour, like a grandfather clock. And when you hold two fingers on your watch face, it tells you the time out loud (by default). To add an hourly tap or chime, launch the Settings app on your Apple Watch, select Sounds & Haptics, tap My Watch at the bottom of the screen, and then tap Notifications.

Activity (analog and digital)

With color-coded activity levels — Move, Exercise, and Stand — these watch faces show your daily activity progress superimposed over a traditional analog clock or beside a larger digital clock. The analog Activity face lets you view your activity rings in the familiar stacked design or as small subdials on the face.

Each option has many available customizable complications to place around the screen:

Activity · Air Quality · Alarm · Audiobooks · Battery · Blood Oxygen (Apple Watch Series 6 and Series 7 only; not available in all regions) · Breathe · Calendar · Camera Remote · Cellular (cellular models only) · Compass · Cycle Tracking · Date · ECG (not available in all regions or for Apple Watch Series 3 or Apple Watch SE) · Elevation (Apple Watch SE, Apple Watch Series 6 and Series 7 only) · Favorite Contacts · Find People · Heart Rate · Home · Mail · Maps ·Medications · Messages · Moon Phase · Music · News · Noise · Now Playing · Phone · Podcasts · Radio · Rain · Reminders · Remote · Shortcuts · Sleep · Stocks · Stopwatch · Sunrise/Sunset · Timer · UV Index · Voice Memos · Walkie-Talkie (not available in all regions) · Weather · Weather Conditions · Wind · Workout · World Clock

Artists

This watch face is one of the newer options for Apple Watch; it changes every time you raise your wrist or tap your Apple Watch screen. Figure 4-1 (left) shows an example of the Artist face. Added in 2022 is a Playtime option within Artists for a more playful way to show the time. Figure 4-1 (right) shows the Playtime option.

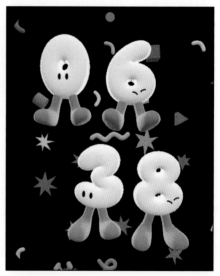

FIGURE 4-1: The Artist watch face (left) and an example of Playtime (right).

Astronomy

Apple says it worked with astrophysicists to create this visually striking watch face, shown in Figure 4-2, which has been updated in 2022 to take advantage of models with larger screens (Apple Watch 7 and 8, and Apple Watch Ultra). The time and date appear at the top of the screen, and you can turn the Digital Crown button to view Earth's rotation, the phases of the moon, and even the entire solar system, all accurately displayed in time. Consider this watch face to be a throwback to the oldest way to tell time: with stars, planets, and our moon.

FIGURE 4-2:
The Astronomy
watch face.

Available complications: Activity · Alarms · Astronomy (Moon Phase) · Audiobooks · Calendar (Today's Date, Your Schedule) · Compass (Compass, Compass/Elevation, Elevation) · Contacts · Controls (Battery) · Heart Rate · Medications · Messages · Music · News · Noise (Sound Levels) · Now Playing · Podcasts · Reminders · Shortcuts · Stocks · Stopwatch · Timer · Weather · Workout · World Clock (Sunrise/Sunset)

Breathe

This watch face, shown in Figure 4-3, was designed to encourage relaxation and mindful breathing — which is now part of the Mindfulness app. At any time in the day, just raise your wrist and follow the rhythm and motion of the Breathe face.

You have three styles to choose from: Classic, Calm, and Focus.

You can add multiple complications to see information in the corners, including: Activity · Alarms · Astronomy (Moon Phase) · Audiobooks · Blood Oxygen · Calculator · Calendar (Today's Date, Your Schedule) · Camera Remote · Compass (Compass, Elevation) · Contacts · Controls (Battery, Cellular) · Cycle Tracking · ECG · Find Devices · Find Items · Find People · Heart Rate · Home · Mail · Maps (Maps, Nearby Transit) · Medications · Messages · Mindfulness · Music · News · Noise (Sound Levels) · Now Playing · Phone · Podcasts · Reminders · Remote · Shortcuts · Sleep · Stocks · Stopwatch · Timer · Tips · Voice Memos · Walkie-Talkie · Weather · Workout · World Clock (Sunrise/Sunset)

California

Debuting in watchOS 6, this Apple Watch face is the first with a California dial (a mix of Roman and Arabic numerals) that you can further customize, making it all Roman, Arabic, Arabic–Indic, or Devanagari. You can select the background color, hands, and complications. See Figure 4-4.

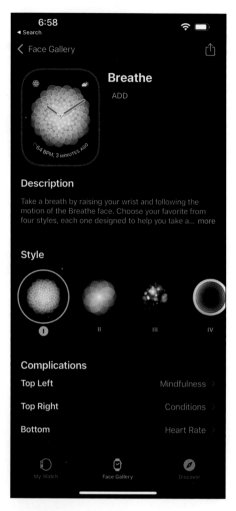

FIGURE 4-3:
Namaste! Here are a few of the Breathe watch face options in the Apple Watch app on the iPhone.

Available complications: Activity · Alarms · Astronomy (Earth, Moon, Moon Phase, Solar, Solar System) · Audiobooks · Blood Oxygen · Calculator · Calendar (Today's Date, Your Schedule) · Camera Remote · Compass (Compass, Elevation) · Contacts · Controls (Battery, Cellular) · Cycle Tracking · ECG · Find Devices · Find Items · Find People · Heart Rate · Home · Mail · Maps (Maps, Nearby Transit) · Medications · Messages · Mindfulness · Monogram · Music · News · Noise (Sound Levels) · Now Playing · Phone · Podcasts · Reminders · Remote · Shortcuts · Sleep · Stocks · Stopwatch · Time (Analog Seconds, Analog Time, Digital Seconds, Digital Time) · Timer · Tips · Voice Memos · Walkie-Talkie · Weather · Workout · World Clock (Sunrise/Sunset)

FIGURE 4-4:
A few options for the California watch face.

Chronograph and Chronograph Pro

Resembling an analog stopwatch, the Chronograph watch face has one main analog clock, with hour, minute, and second hands, and two additional smaller hands: one for total time and a second for lap times, as shown in Figure 4-5. These secondary faces are very customizable too. You can also choose to place additional information in the four corners, as explained in greater depth in "Differentiating Between Customizations and Complications" later in this chapter. Chronograph Pro lets you record time on scales of 60, 30, 6, or 3 seconds, or select the tachymeter time scale to measure speed between distances.

FIGURE 4-5:
The Chronograph watch face.

Available complications: Activity · Alarm · Audiobooks · Battery · Blood Oxygen (Apple Watch Series 6 and Series 7 only; not available in all regions) · Calculator · Calendar · Camera Remote · Cellular (cellular models only) · Compass · Cycle Tracking · Date · ECG (not available in all regions or for Apple Watch Series 3 or Apple Watch SE) · Elevation (Apple Watch SE, Apple Watch Series 6 and Series 7 only) · Favorite Contacts · Find People · Heart Rate · Home · Mail · Maps · Medications · Messages · Mindfulness · Moon Phase · Music · News · Noise · Phone · Podcasts · Radio · Rain · Reminders · Remote · Shortcuts · Stocks · Sunrise/Sunset · Timer · UV Index · Voice Memos · Walkie-Talkie · Weather · Weather Conditions · Wind · Workout · World Clock (Sunrise/Sunset)

Color

By twisting the Digital Crown button, you can choose a watch-face color that suits your outfit, style, or mood. This classic analog face can be as minimalist or as busy as you like based on the number of complications you select. Figure 4-6 shows the Color watch face.

FIGURE 4-6: The Color watch face.

Available complications (for Circular and Dial styles only): Activity · Alarms · Astronomy (Moon Phase) · Audiobooks · Blood Oxygen · Calculator · Calendar (Today's Date, Your Schedule) · Camera Remote · Compass (Compass, Elevation)

· Contacts · Controls (Battery, Cellular) · Cycle Tracking · ECG · Find Devices · Find Items · Find People · Heart Rate · Home · Mail · Maps (Maps, Nearby Transit) · Medications · Messages · Mindfulness · Music · News · Noise (Sound Levels) · Phone · Podcasts · Reminders · Remote · Shortcuts · Sleep · Stocks · Stopwatch · Timer · Tips · Voice Memos · Walkie-Talkie · Weather · Workout · World Clock (Sunrise/Sunset)

Contour

Designed for the larger Apple Watch Series 7 and Apple Watch Series 8 — and looking stellar on the biggest in the family, the Apple Watch Ultra — the Contour face places the numbers right up against the edges of the watch (as shown in Figure 4-7) and fluidly animates throughout the day while emphasizing the current hour.

Pro tip: Twist the Digital Crown to see the numbers animate along the edges of the watch face!

Available complications: Activity · Alarms · Astronomy (Earth, Moon, Solar, Solar System) · Audiobooks · Blood Oxygen · Calculator · Calendar (Today's Date) · Camera Remote · Compass (Compass, Elevation) · Contacts · Controls (Battery, Cellular) · Cycle Tracking · ECG · Find Devices · Find Items · Find People · Heart Rate · Home · Mail · Maps (Maps, Nearby Transit) · Medications · Messages · Mindfulness · Monogram · Music · News · Noise (Sound Levels) · Phone · Podcasts · Reminders · Remote · Shortcuts · Sleep · Stocks · Stopwatch · Time (Analog Seconds, Analog Time, Digital Seconds, Digital Time) · Timer · Tips · Voice Memos · Walkie-Talkie · Weather · Workout · World Clock (Sunrise/Sunset)

FIGURE 4-7:
Contour is one of the newest watch faces, boasting a stylish analog design, and was created exclusively for the launch of Apple Watch Series 7 in 2021.

Count Up

Track elapsed time in this watch face, which is available only for Apple Watch Series 4 and later or Apple Watch SE. See Figure 4-8.

FIGURE 4-8:
The Count Up watch face.

Available complications (for Circular and Dial styles only): Activity · Alarms · Astronomy (Moon) · Audiobooks · Blood Oxygen · Calculator · Calendar (Today's Date, Your Schedule) · Camera Remote · Compass (Compass, Elevation) · Contacts · Controls (Battery, Cellular) · Cycle Tracking · ECG · Find Devices · Find Items · Find People · Heart Rate · Home · Mail · Maps (Maps, Nearby Transit) · Medications · Messages · Mindfulness · Monogram · Music · News · Noise (Sound Levels) · Phone · Podcasts · Reminders · Remote · Shortcuts · Sleep · Stocks · Stopwatch · Timer · Tips · Voice Memos · Walkie-Talkie · Weather · Workout · World Clock (Sunrise/Sunset)

Explorer

Quite simply, you can find the Explorer watch analog face (see Figure 4-9) only on Apple Watch with GPS + cellular. It prominently features green dots, which indicate cellular signal strength in your area. You can customize the color hands (red, red/white, red/gray); choose among four styles; and add optional complications such as Activity, Walkie-Talkie, News, and Moon Phase.

FIGURE 4-9:
On Apple Watch with GPS + Cellular, the Explorer watch face has green dots that show you the strength of the cellular connection near you.

Available complications: Activity · Alarms · Astronomy (Moon Phase) · Audiobooks · Blood Oxygen · Calculator · Calendar (Today's Date, Your Schedule) · Camera Remote · Compass (Compass, Compass/Elevation, Elevation) · Contacts · Controls (Battery, Cellular) · Cycle Tracking · ECG · Find Devices · Find Items · Find People · Heart Rate · Home · Mail · Maps (Maps, Nearby Transit) · Medications · Messages · Mindfulness · Monogram · Music · News · Noise (Sound Levels) · Now Playing · Phone · Podcasts · Reminders · Remote · Shortcuts · Sleep · Stocks · Stopwatch · Timer · Tips · Voice Memos · Walkie-Talkie · Weather · Workout · World Clock (Sunrise/Sunset)

Fire and Water

This watch face (see Figure 4-10) brings two earthly elements to Apple Watch! Apple says that each animated film was shot in a custom model, allowing fire and water to define the edges of the face and interact with the dial. This watch face animates whenever you raise your wrist or tap the display. You can choose Fire, Water, or Fire and Water. For Apple Watch Series 4 and later, you can also choose full-screen or circular.

Several dozen complications are available (circular style only): Activity · Alarms · Astronomy (Moon Phase) · Audiobooks · Blood Oxygen · Calculator · Calendar (Today's Date, Your Schedule) · Camera Remote · Compass (Compass, Compass/ Elevation, Elevation) · Contacts · Controls (Battery, Cellular) · Cycle Tracking

· ECG · Find Devices · Find Items · Find People · Heart Rate · Home · Mail · Maps (Maps, Nearby Transit) · Medications · Messages · Mindfulness · Music · News · Noise (Sound Levels) · Now Playing · Phone · Podcasts · Reminders · Remote · Shortcuts · Sleep · Stocks · Stopwatch · Timer · Tips · Voice Memos · Walkie-Talkie · Weather · Workout · World Clock (Sunrise/Sunset)

FIGURE 4-10:
Choose Fire, Water, or Fire and Water together for the animated Fire and Water watch face.

GMT

Available only for Apple Watch Series 4 and later and Apple Watch SE, the GMT face (shown in Figure 4-11) offers two dials. The 12-hour inner dial displays the local time, and the outer 24-hour dial lets you track a second time zone.

Available complications: Activity · Alarms · Astronomy (Moon) · Audiobooks · Blood Oxygen · Calculator · Calendar (Today's Date, Your Schedule) · Camera Remote · Compass (Compass, Elevation) · Contacts · Controls (Battery, Cellular) · Cycle Tracking · ECG · Find Devices · Find Items · Find People · Heart Rate · Home · Mail · Maps (Maps, Nearby Transit) · Medications · Messages · Mindfulness · Music · News · Noise (Sound Levels) · Phone · Podcasts · Reminders · Remote · Shortcuts · Sleep · Stocks · Stopwatch · Timer · Tips · Voice Memos · Walkie-Talkie · Weather · Workout · World Clock (Sunrise/Sunset)

FIGURE 4-11:
GMT is one
of the newer
watch-face
options.

Gradient

As shown in Figure 4-12, the Gradient watch face is simple and elegant. You can choose among three styles of gradients (full-screen or circular), each of which moves with the time in different ways. You can also add complications to the top left, top right, bottom left, bottom right, or top middle of the watch face.

Available complications: Activity · Alarms · Astronomy (Moon, Moon Phase) · Audiobooks · Blood Oxygen · Calculator · Calendar (Today's Date, Your Schedule) · Camera Remote · Compass (Compass, Compass/Elevation, Elevation) · Contacts · Controls (Battery, Cellular) · Cycle Tracking · ECG · Find Devices · Find Items · Find People · Heart Rate · Home · Mail · Maps (Maps, Nearby Transit) · Medications · Messages · Mindfulness · Music · News · Noise (Sound Levels) · Now Playing · Phone · Podcasts · Reminders · Remote · Shortcuts · Sleep · Stocks · Stopwatch · Timer · Tips · Voice Memos · Walkie-Talkie · Weather · Workout · World Clock (Sunrise/Sunset)

Infograph, Infograph Modular

Available in Apple Watch Series 4 and later, the Infograph watch face (see Figure 4-13) features up to eight rich, full-color complications and subdials, whereas the Infograph Modular face has up to six rich, full-color complications. You can choose any of dozens of colors for the outside lining of the face.

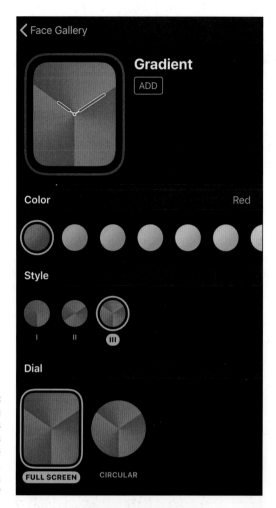

FIGURE 4-12:
This watch face features gradients that move with the time. Which color is for you?

FIGURE 4-13: The Infograph (left) and Infograph Modular (right) watch faces offer a ton of customizable information you can see at a glance.

Available complications: Activity · Alarms · Astronomy (Earth, Moon, Solar, Solar System) · Audiobooks · Blood Oxygen · Calculator · Calendar (Today's Date, Your Schedule) · Camera Remote · Compass (Compass, Elevation) · Contacts · Controls (Battery, Cellular) · Cycle Tracking · ECG · Find Devices · Find Items · Find People · Heart Rate · Home · Mail · Maps (Maps, Nearby Transit) · Medications · Messages · Mindfulness · Music · News · Noise (Sound Levels) · Now Playing · Phone · Podcasts · Reminders · Remote · Shortcuts · Sleep · Stocks · Stopwatch · Time (Analog Seconds, Analog Time, Digital Seconds, Digital Time) · Timer · Tips · Voice Memos · Walkie-Talkie · Weather · Workout · World Clock (Sunrise/Sunset)

Kaleidoscope

As you may surmise, with a watch face called Kaleidoscope (shown in Figure 4-14), you can select a photo to create a watch face with evolving patterns of shapes and colors; simply turn the Digital Crown button to change the pattern. You can choose a custom photo or select one of the many photos in the Apple Watch app for the iPhone (Mirror, Flower, Graphic, and more).

FIGURE 4-14: Trippy, man! The Kaleidoscope watch face for Apple Watch features evolving patterns and shapes, with lots of options available.

You can choose among multiple styles (Facet, Radial, and Rosette) and customizable complications: Activity · Alarms · Astronomy (Moon Phase) · Audiobooks · Blood Oxygen · Calculator · Calendar (Today's Date, Your Schedule) · Camera

Remote · Compass (Compass, Compass/Elevation, Elevation) · Contacts · Controls (Battery, Cellular) · Cycle Tracking · ECG · Find Devices · Find Items · Find People · Heart Rate · Home · Mail · Maps (Maps, Nearby Transit) · Medications · Messages · Mindfulness · Music · News · Noise (Sound Levels) · Now Playing · Phone · Podcasts · Reminders · Remote · Shortcuts · Sleep · Stocks · Stopwatch · Time (Digital Time) · Timer · Tips · Voice Memos · Walkie-Talkie · Weather · Workout · World Clock (Sunrise/Sunset)

Liquid Metal

The Liquid Metal watch face (see Figure 4-15) animates whenever you raise your wrist or tap the display. Seriously, this watch face looks super-cool! For Apple Watch Series 4 and later, you can also choose among four colors and two styles (full-screen and circular).

FIGURE 4-15: Liquid Metal watch face options add a high-tech look to your Apple Watch.

Available complications (circular style only): Activity · Alarms · Astronomy (Moon Phase) · Audiobooks · Blood Oxygen · Calculator · Calendar (Today's Date, Your Schedule) · Camera Remote · Compass (Compass, Compass/Elevation, Elevation) · Contacts · Controls (Battery, Cellular) · Cycle Tracking · ECG · Find Devices · Find Items · Find People · Heart Rate · Home · Mail · Maps (Maps, Nearby Transit) · Medications · Messages · Mindfulness · Music · News · Noise (Sound Levels)

· Now Playing · Phone · Podcasts · Reminders · Remote · Shortcuts · Sleep · Stocks · Stopwatch · Timer · Tips · Voice Memos · Walkie-Talkie · Weather · Workout · World Clock (Sunrise/Sunset)

Lunar

As the name suggests, Lunar celebrates calendar timekeeping using phases of the moon. You can choose from three different calendars: Chinese, Hebrew, and Islamic. (See Figure 4-16.)

Available complications: Activity · Alarms · Astronomy (Moon) · Audiobooks · Blood Oxygen · Calculator · Calendar (Today's Date) · Camera Remote · Compass (Compass, Elevation) · Contacts · Controls (Battery, Cellular) · Cycle Tracking · ECG · Find Devices · Find Items · Find People · Heart Rate · Home · Mail · Maps (Maps, Nearby Transit) · Medications · Messages · Mindfulness · Monogram · Music · News · Noise (Sound Levels) · Phone · Podcasts · Reminders · Remote · Shortcuts · Sleep · Stocks · Stopwatch · Time (Analog Seconds, Analog Time, Digital Seconds, Digital Time) · Timer · Tips · Voice Memos · Walkie-Talkie · Weather · Workout · World Clock (Sunrise/Sunset)

FIGURE 4-16: An example of the Lunar clock face, first introduced in watchOS9.

Memoji

A new watch-face option, Memoji, displays the Memoji you created in the app, which could be a fun cartoony representation of you, or you can use other Memoji characters (see Figure 4-17).

Available complications: Activity · Alarms · Astronomy (Moon Phase) · Audiobooks · Calendar (Today's Date, Your Schedule) · Compass (Compass, Compass/Elevation, Elevation) · Controls (Battery) · Heart Rate · Medications · Messages · Music · News · Noise (Sound Levels) · Now Playing · Podcasts · Reminders · Shortcuts · Stocks · Stopwatch · Timer · Weather · Workout · World Clock (Sunrise/Sunset)

Meridian

Enjoy this classic and full-screen analog watch face, available in black or white and with four customizable subdials. It's available only for Apple Watch Series 4 and later and Apple Watch SE. See Figure 4-18.

Available complications: Activity · Alarms · Astronomy (Earth, Moon, Solar, Solar System) · Audiobooks · Blood Oxygen · Calculator · Calendar (Today's Date) · Camera Remote · Compass (Compass, Elevation) · Contacts · Controls (Battery, Cellular) · Cycle Tracking · ECG · Find Devices · Find Items · Find People · Heart Rate · Home · Mail · Maps (Maps, Nearby Transit) · Medications · Messages · Mindfulness · Monogram · Music · News · Noise (Sound Levels) · Phone · Podcasts

· Reminders · Remote · Shortcuts · Sleep · Stocks · Stopwatch · Time (Analog Seconds, Analog Time, Digital Seconds, Digital Time) · Timer · Tips · Voice Memos · Walkie-Talkie · Weather · Workout · World Clock (Sunrise/Sunset)

FIGURE 4-18:
Meridian is an elegant but functional watch face.

Metropolitan

Metropolitan is a classic, type-driven watch face. The custom-designed numbers dynamically change in style and weight as you turn the Digital Crown. The numbers tend to stretch out the more you twist the crown, for example, so you can keep what you like. You can select the background color and up to four complications with the circular dial. (See Figure 4-19.)

Available complications: Activity · Alarms · Astronomy (Moon) · Audiobooks · Blood Oxygen · Calculator · Calendar (Today's Date) · Camera Remote · Compass (Compass, Elevation) · Contacts · Controls (Battery, Cellular) · Cycle Tracking · ECG · Find Devices · Find Items · Find People · Heart Rate · Home · Mail · Maps (Maps, Nearby Transit) · Medications · Messages · Mindfulness · Monogram · Music · News · Noise (Sound Levels) · Phone · Podcasts · Reminders · Remote · Shortcuts · Sleep · Stocks · Stopwatch · Time (Analog Seconds, Analog Time, Digital Seconds, Digital Time) · Timer · Tips · Voice Memos · Walkie-Talkie · Weather · Workout · World Clock (Sunrise/Sunset)

FIGURE 4-19: First available in the Fall 2022, Metropolitan is a classic analog watch face, but with support for several complications.

Mickey Mouse, Minnie Mouse

The classic Mickey Mouse and Minnie Mouse analog watches (see Figure 4-20) have been reinvented for Apple Watch. Mickey's or Minnie's arms move around the dial as they point to the correct hour and minute, while the character taps a foot every second. You can choose various customizations. In addition, you can choose whether your character speaks the time! To enable that feature, open the Apple Watch app on your iPhone and tap My Watch ⇨ Sounds & Haptics ⇨ Tap to Speak Time. You won't hear the voice if your Apple Watch is in Silent mode.

Available complications: Activity · Alarms · Astronomy (Moon Phase) · Audiobooks · Blood Oxygen · Calculator · Calendar (Today's Date, Your Schedule) · Camera Remote · Compass (Compass, Compass/Elevation, Elevation) · Contacts · Controls (Battery, Cellular) · ECG · Find Devices · Find Items · Find People · Heart Rate · Home · Mail · Maps (Maps, Nearby Transit) · Medications · Messages · Mindfulness · Music · News · Noise (Sound Levels) · Now Playing · Phone · Podcasts · Reminders · Shortcuts · Sleep · Stocks · Stopwatch · Timer · Tips · Voice Memos · Walkie-Talkie · Weather · Workout · World Clock (Sunrise/Sunset)

OH, MICKEY, YOU'RE SO FINE

Apple Watch isn't the only product that Apple has partnered with Mickey Mouse. Apple launched a new iPod nano back in the fall of 2011, and with it came 16 new clock faces, including ones featuring the beloved Disney characters Mickey Mouse and Minnie Mouse.

Did you know that the original Mickey Mouse watch was an instant hit when it debuted back in 1933? In fact, it saved the cash-strapped Ingersoll company from bankruptcy. (As it did for many other companies, the Great Depression hurt Ingersoll's business considerably.) The Mickey Mouse watch originally sold for $2.98 at Macy's and $2.69 at Sears, Roebuck & Company.

FIGURE 4-20: The Mickey Mouse watch face.

Modular/Modular Compact/Modular Duo

As the name suggests, this digital watch face allows the greatest number of complications of all watch faces and gives you a ton of extra information at a glance. The interface can be as clean or cluttered as you see fit. With Modular Compact, shown in Figure 4-21, you have even more flexibility when adding complications, colors, and choosing between analog and digital time. Figure 4-22 shows Modular Duo, designed only for use with the larger Apple Watch Series 7, Apple Watch Series 8 and Apple Watch Ultra, all of which support even more data at a glance.

FIGURE 4-21:
The Modular
Compact watch
face.

Available complications: Activity · Alarms · Astronomy (Earth, Moon, Solar, Solar System) · Audiobooks · Blood Oxygen · Calculator · Calendar (Today's Date, Your Schedule) · Camera Remote · Compass (Compass, Elevation) · Contacts · Controls (Battery, Cellular) · Cycle Tracking · ECG · Find Devices · Find Items · Find People · Heart Rate · Home · Mail · Maps (Maps, Nearby Transit) · Medications · Messages · Mindfulness · Music · News · Noise (Sound Levels) · Now Playing · Phone · Podcasts · Reminders · Remote · Shortcuts · Sleep · Stocks · Stopwatch · Time (Analog Seconds, Analog Time, Digital Seconds, Digital Time) · Timer · Tips · Voice Memos · Walkie-Talkie · Weather · Workout · World Clock (Sunrise/Sunset)

REMEMBER

Available on Modular Compact, Modular, and X-Large faces, you can now customize your watch face with a wide range of colors and gradients to match your mood — from aqua to zinc! Simply tap the Edit button when on a supported watch face and customize away! (See Figure 4-23.)

Motion

Apple Watch fans are going to love this one. This watch face displays a different animated image every time you raise your wrist. Based on the theme you choose, raise your wrist, and you may see a butterfly slowly fluttering its wings (as shown in Figure 4-24) or a flower blooming.

You can choose among many colors, as well as complications: Activity · Alarms · Astronomy (Moon Phase) · Audiobooks · Calendar (Today's Date, Your Schedule) · Compass (Compass, Compass/Elevation, Elevation) · Controls (Battery) · Heart Rate · Medications · Messages · Music · News · Noise (Sound Levels) · Now Playing · Podcasts · Reminders · Shortcuts · Stocks · Stopwatch · Timer · Weather · Workout · World Clock (Sunrise/Sunset)

TECHNICAL STUFF

Apple always goes above and beyond. Some objects for the Motion watch face were video-recorded, such as the jellyfish (at 300 frames per second); others, such as the blooming flowers, were created by using stop-motion, time-lapse photos. Apple says a single flower took more than 285 hours and 24,000 shots to photograph.

FIGURE 4-22:
Taking advantage of the 20 percent larger screen in Apple Watch Series 7 and Apple Watch Series 8 (and even larger for Apple Watch Ultra), the Modular Duo face offers two large areas for complications.

Numerals, Numerals Duo/Numerals Mono

Fancy-schmancy! The Numerals watch face displays digital and analog time, with many customizable elements. The Numerals Duo shows a large digital display in different colors (see Figure 4-25). Similar to Numerals, Numerals Mono face offers a (larger) digital and analog hybrid, with different colors to choose from. Why not change the colors and styles to match your outfit?

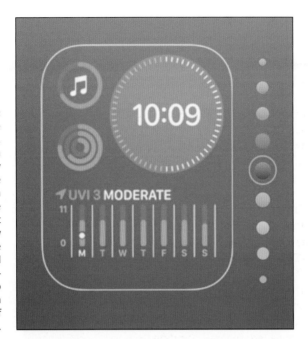

FIGURE 4-23:
With Apple
Watch,
customization
is key! Not only
do you have
several watch
faces to choose
from, but
you can now
choose more
background
colors —
perhaps to
match with
your outfit of
the day.

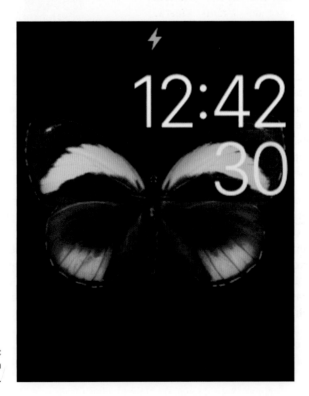

FIGURE 4-24:
The Motion
watch face.

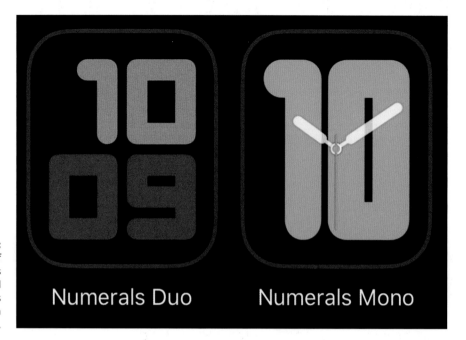

FIGURE 4-25:
Examples of
the Numerals
Duo and
Numerals
Mono watch
faces.

Available complications for Numerals (not available on Numerals Duo or Numerals Mono): Activity · Alarms · Astronomy (Moon Phase) · Audiobooks · Blood Oxygen · Calculator · Calendar (Today's Date, Your Schedule) · Compass (Compass, Compass/Elevation, Elevation) · Controls (Battery) · Heart Rate · Medications · Messages · Music · News · Noise (Sound Levels) · Now Playing · Podcasts · Reminders · Shortcuts · Stocks · Stopwatch · Timer · Tips · Weather · Workout · World Clock (Sunrise/Sunset)

Photos

What better way to personalize your watch than by having a photo on it as your clock? With the aptly named Photos face (see Figure 4-26), you can display a photo (or multiple photos) from your iPhone or iCloud gallery. In fact, set things up so that a new photo appears every time you raise your wrist or tap your display. Choose images from a synced album, recent Memories, or up to ten custom photos. To add more photos, firmly press the watch face, swipe all the way to the right, tap the New button (+), and then tap Photos. Or, in the Photos app on your Apple Watch, press any photo and then tap Create Watch Face.

Available complications: Activity · Alarms · Astronomy (Moon Phase) · Audiobooks · Calendar (Today's Date, Your Schedule) · Compass (Compass, Compass/Elevation, Elevation) · Controls (Battery) · Heart Rate · Medications · Messages · Music · News · Noise (Sound Levels) · Now Playing · Podcasts · Reminders · Shortcuts · Stocks · Stopwatch · Timer · Weather · Workout · World Clock (Sunrise/Sunset)

FIGURE 4-26:
One of the
most fun
watch-face
options is
Photos,
because you
can see a
different photo
every time you
lift your wrist!

Portraits

According to Apple, the Photos face (preceding section) is the most popular Apple Watch face, so watchOS 8 introduces an extension called Portraits. This watch face brings to life stunning Portrait photos shot on the iPhone (you know — the ones with that blurred-background depth-of-field effect) and adds the date and time, as you see in Figure 4-27. Beginning with watchOS9, you can also put pictures of your dog or cat on your watch face, plus you can now add a tint to the background layers of a photo.

Available complications: Activity · Alarms · Astronomy (Moon Phase) · Audiobooks · Calendar (Today's Date, Your Schedule) · Compass (Compass, Compass/Elevation, Elevation) · Controls (Battery) · Heart Rate · Medications · Messages · Music · News · Noise (Sound Levels) · Now Playing · Podcasts · Reminders · Shortcuts · Stocks · Stopwatch · Timer · Weather · Workout · World Clock (Sunrise/Sunset)

TIP

Turn the Digital Crown to zoom in on the face.

Pride Analog, Pride Digital, Pride Woven

Inspired by the rainbow flag, a symbol of LGBTQ (lesbian, gay, bisexual, transgender, and queer/questioning), this watch face offers multiple colored ribbons that move if you tap the display (see Figure 4-28). You can choose an analog or digital clock.

TIP

Did you know you can add the Pride look to other watch faces? Open the Apple Watch app on your iPhone and select Face Gallery. Then, in the Pride section, choose multicolor options for the California, Numerals Duo, Numerals Mono, and Gradient watch faces.

Available complications: Activity · Alarms · Astronomy (Moon Phase) · Audiobooks · Blood Oxygen · Calculator · Calendar (Today's Date, Your Schedule) · Camera Remote · Compass (Compass, Elevation) · Contacts · Controls (Battery, Cellular) · Cycle Tracking · ECG · Find Devices · Find Items · Find People · Heart Rate · Home · Mail · Maps (Maps, Nearby Transit) · Medications · Messages · Mindfulness · Music · News · Noise (Sound Levels) · Phone · Podcasts · Reminders · Remote · Shortcuts · Sleep · Stocks · Stopwatch · Timer · Tips · Voice Memos · Walkie-Talkie · Weather · Workout · World Clock (Sunrise/Sunset)

FIGURE 4-27:
The new Portraits face lets you choose a photo from your iPhone captured in Portraits mode to create a unique and sentimental watch face.

Simple

The most minimalist of all Apple Watch faces, Simple — as the name implies — offers a straightforward yet elegant face with analog hands for minute, hour, and second, as shown in Figure 4-29. A single number tells you the day of the month. As with all other watch faces, you can adjust the amount of detail with the Digital Crown button.

FIGURE 4-28:
Wear your pride on your wrist. The Pride watch face shows multicolored ribbons on your Apple Watch clock face.

FIGURE 4-29:
The Simple
watch face.

Available complications: Activity · Alarms · Astronomy (Moon) · Audiobooks · Blood Oxygen · Calculator · Calendar (Today's Date, Your Schedule) · Camera Remote · Compass (Compass, Elevation) · Contacts · Controls (Battery, Cellular) · Cycle Tracking · ECG · Find Devices · Find Items · Find People · Heart Rate · Home · Mail · Maps (Maps, Nearby Transit) · Medications · Messages · Mindfulness · Music · News · Noise (Sound Levels) · Phone · Podcasts · Reminders · Remote · Shortcuts · Sleep · Stocks · Stopwatch · Timer · Tips · Voice Memos · Walkie-Talkie · Weather · Workout · World Clock (Sunrise/Sunset)

Siri

Now your favorite personal assistant can be part of your clock face. With the Siri watch face, your artificial-intelligence (AI) pal takes a look at your day and displays information that's timely, relevant, and helpful (as shown in Figure 4-30). This information could be traffic conditions on your way home from work, the day and time of your next calendar appointment, or the latest score of your favorite sports team. Tap for a deeper dive, or twist the Digital Crown button to scroll through your day.

Available complications: Activity · Alarms · Astronomy (Moon Phase) · Audiobooks · Blood Oxygen · Calculator · Calendar (Today's Date, Your Schedule) · Camera Remote · Compass (Compass, Elevation) · Contacts · Controls (Battery, Cellular) · Cycle Tracking · ECG · Find Devices · Find Items · Find People · Heart Rate · Home · Mail · Maps (Maps, Nearby Transit) · Medications · Messages · Mindfulness · Music · News · Noise (Sound Levels) · Phone · Podcasts · Reminders · Remote · Shortcuts · Siri · Sleep · Stocks · Stopwatch · Timer · Tips · Voice Memos · Walkie-Talkie · Weather · Workout · World Clock (Sunrise/Sunset)

Solar Dial and Solar Graph

Along with showing you the time in analog or digital format, Solar Dial features a 24-hour circular dial that tracks the sun's position in the sky, as shown in Figure 4-31. Twist the Digital Crown button to trace the sun's arc over the course of the day. Solar Dial, on the other hand, displays a 24-hour circular dial that tracks the sun, as well as an analog or digital dial that moves opposite to the sun's path; you can tap the watch face to see the day's length. This watch face is available only for Apple Watch Series 4 and later and Apple Watch SE.

Turn the Digital Crown to move through the day's solar events.

FIGURE 4-31:
The Solar Dial
watch face.

Available complications for Solar: Activity · Alarms · Astronomy (Moon) · Audiobooks · Blood Oxygen · Calculator · Calendar (Today's Date, Your Schedule) · Camera Remote · Compass (Compass, Elevation) · Contacts · Controls (Battery, Cellular) · Cycle Tracking · ECG · Find Devices · Find Items · Find People · Heart Rate · Home · Mail · Maps (Maps, Nearby Transit) · Medications · Messages · Mindfulness · Music · News · Noise (Sound Levels) · Phone · Podcasts · Reminders · Remote · Shortcuts · Sleep · Stocks · Stopwatch · Timer · Tips · Voice Memos · Walkie-Talkie · Weather · Workout · World Clock (Sunrise/Sunset)

Available complications for Solar Graph: Activity · Alarms · Astronomy (Moon Phase) · Audiobooks · Calendar (Today's Date, Your Schedule) · Camera Remote · Compass (Compass, Compass/Elevation, Elevation) · Controls (Battery) · Heart Rate · Messages · Music · News · Noise (Sound Levels) · Now Playing · Podcasts · Reminders · Shortcuts · Stocks · Stopwatch · Timer · Weather · Workout · World Clock (Sunrise/Sunset)

Stripes

This newer watch-face option lets you select the number of stripes you want, as well as colors and rotation. Stripes (see Figure 4-32) is only for Apple Watch Series 4 and later and Apple Watch SE.

Available complications (circular style only): Activity · Alarms · Astronomy (Moon) · Audiobooks · Blood Oxygen · Calculator · Calendar (Today's Date, Your Schedule) · Camera Remote · Compass (Compass, Elevation) · Contacts · Controls (Battery, Cellular) · Cycle Tracking · ECG · Find Devices · Find Items · Find People · Heart Rate · Home · Mail · Maps (Maps, Nearby Transit) · Medications · Messages · Mindfulness · Music · News · Noise (Sound Levels) · Phone · Podcasts · Reminders · Remote · Shortcuts · Sleep · Stocks · Stopwatch · Timer · Tips · Voice Memos · Walkie-Talkie · Weather · Workout · World Clock (Sunrise/Sunset)

TimeLapse

Who doesn't love time-lapse videos? Now you can see one — or a cityscape or natural setting — every time you raise your wrist to glance at the time (see Figure 4-33). You can choose which Apple TimeLapse video to see: Mack Lake (in California), New York, Hong Kong, London, Paris, or Shanghai.

Available complications: Activity · Alarms · Astronomy (Moon Phase) · Audiobooks · Calendar (Today's Date, Your Schedule) · Compass (Compass, Compass/Elevation, Elevation) · Controls (Battery) · Heart Rate · Medications · Messages · Music · News · Noise (Sound Levels) · Now Playing · Podcasts · Reminders · Shortcuts · Stocks · Stopwatch · Timer · Weather · Workout · World Clock (Sunrise/Sunset)

Toy Story

If animated films starring *Toy Story* characters are more your speed than classic Mickey Mouse cartoons, Apple Watch has you covered just the same (see Figure 4-34). The Toy Story watch face lets you see animated *Toy Story* characters — Woody, Buzz, Jessie, and Toy Box — whenever you raise your wrist to look at the time.

Available complications: Activity · Alarms · Astronomy (Moon Phase) · Audiobooks · Calendar (Today's Date, Your Schedule) · Compass (Compass, Compass/Elevation, Elevation) · Controls (Battery) · Heart Rate · Medications · Messages · Music · News · Noise (Sound Levels) · Now Playing · Podcasts · Reminders · Shortcuts · Stocks · Stopwatch · Timer · Weather · Workout · World Clock (Sunrise/Sunset)

Typograph

Featuring three custom fonts, including Roman numerals, this watch face (see Figure 4-35) is only for Apple Watch Series 6, Series 7, and Apple Watch SE.

Available complications: Date · Monogram · Stopwatch · Digital Time · Timer

Unity

Unity was inspired by the red and green colors of the Pan-African Flag. The shapes change as you move.

Available complications: Activity · Alarms · Astronomy (Moon Phase) · Audiobooks · Calendar (Today's Date, Your Schedule) · Compass (Compass, Compass/Elevation, Elevation) · Controls (Battery) · Heart Rate · Medications · Messages · Music · News · Noise (Sound Levels) · Now Playing · Podcasts · Reminders · Shortcuts · Stocks · Stopwatch · Timer · Weather · Workout · World Clock (Sunrise/Sunset)

Utility

The most straightforward and practical face of the bunch, Utility shows a classic analog watch face with plenty of space in the corners for extra information, such as World Clock, a timer, or a calendar appointment (see Figure 4-36).

FIGURE 4-36:
The Utility watch face.

Available complications: Activity · Alarms · Astronomy (Moon Phase) · Audiobooks · Blood Oxygen · Calculator · Calendar (Today's Date, Your Schedule) · Camera Remote · Compass (Compass, Elevation) · Contacts · Controls (Battery,

Cellular) · Cycle Tracking · ECG · Find Devices · Find Items · Find People · Heart Rate · Home · Mail · Maps (Maps, Nearby Transit) · Medications · Messages · Mindfulness · Music · News · Noise (Sound Levels) · Phone · Podcasts · Reminders · Remote · Shortcuts · Sleep · Stocks · Stopwatch · Timer · Tips · Voice Memos · Walkie-Talkie · Weather · Workout · World Clock (Sunrise/Sunset)

Vapor

The Vapor watch face (see Figure 4-37) animates . . . well, vapor whenever you raise your wrist or tap the display. As for how it was created, Apple says that each of the four films was shot at thousands of frames per second as a mortar fired colors into a custom chamber. Cool! Choose the color and style that suit your taste.

FIGURE 4-37: The mysterious, sophisticated Vapor watch face for Apple Watch.

Available complications (circular style only): Activity · Alarms · Astronomy (Moon Phase) · Audiobooks · Blood Oxygen · Calculator · Calendar (Today's Date, Your Schedule) · Camera Remote · Compass (Compass, Compass/Elevation, Elevation) · Contacts · Controls (Battery, Cellular) · Cycle Tracking · ECG · Find Devices · Find Items · Find People · Heart Rate · Home · Mail · Maps (Maps, Nearby Transit) · Medications · Messages · Mindfulness · Music · News · Noise (Sound Levels) · Now Playing · Phone · Podcasts · Reminders · Remote · Shortcuts · Sleep · Stocks · Stopwatch · Timer · Tips · Voice Memos · Walkie-Talkie · Weather · Workout · World Clock (Sunrise/Sunset)

World Time

Perfect for travelers and inspired by heritage watches, World Time, as the name suggests, tracks the time in time zones around the world — 24 of them, in fact — and can be used on Apple Watch Series 4 and later. While pictured in Figure 4-38 on Apple Watch Series 8, it works on Apple Watch Series 4, and newer.

FIGURE 4-38:
One of the newest watch faces, World Time tracks the time in 24 time zones around a double dial.

Available complications: Activity · Alarms · Astronomy (Moon Phase) · Audiobooks · Blood Oxygen · Calculator · Calendar (Today's Date, Your Schedule) · Camera Remote · Compass (Compass, Elevation) · Contacts · Controls (Battery, Cellular) · Cycle Tracking · ECG · Find Devices · Find Items · Find People · Heart Rate · Home · Mail · Maps (Maps, Nearby Transit) · Medications · Messages · Mindfulness · Music · News · Noise (Sound Levels) · Phone · Podcasts · Reminders · Remote · Shortcuts · Sleep · Stocks · Stopwatch · Timer · Tips · Voice Memos · Walkie-Talkie · Weather · Workout · World Clock (Sunrise/Sunset)

X-Large

As you might expect with a name like X-Large, this watch face shows a very large digital clock with the hour at the top of the screen and the date at the bottom, as shown in Figure 4-39. Although the background is black, you can adjust the color to your liking — perhaps to match an outfit you're wearing. You can also add complications, which take up part of the screen too.

FIGURE 4-39:
The X-Large watch face.

Available complications: Activity · Alarms · Astronomy (Earth, Moon, Solar, Solar System) · Audiobooks · Blood Oxygen · Calculator · Calendar (Today's Date, Your Schedule) · Camera Remote · Compass (Compass, Elevation) · Contacts · Controls (Battery, Cellular) · Cycle Tracking · ECG · Find My (Find People) · Heart Rate · Home · Mail · Maps (Maps, Nearby Transit) · Medications · Messages · Mindfulness · Music · News · Noise (Sound Levels) · Phone · Podcasts · Reminders · Remote · Shortcuts · Sleep · Stocks · Stopwatch · Timer · Tips · Voice Memos · Walkie-Talkie · Weather · Workout · World Clock (Sunrise/Sunset)

Choosing Among the Various Watch Faces

One of the first things you may do with Apple Watch is choose one of the many, many watch faces provided and then customize it to your liking — perhaps with complications. And it's super-easy to change faces and make those optional changes on the watch itself.

The first step is loading a few desirable faces onto the watch, which you can handle wirelessly via your iPhone! To load watch faces, follow these easy steps:

1. **Open the Apple Watch app on your iPhone.**

2. **Tap Face Gallery at the bottom of the screen.**

3. **Scroll up and down the screen to see multiple watch faces, as well as style options and complications (see Figure 4-40).**

4. **If any watch face looks interesting to you, tap the Add button.**

   Your iPhone wirelessly beams the watch face over to Apple Watch like magic.

That's it!

Now you have a selection of clock faces to choose from on the Apple Watch itself. To select a specific watch face for your Apple Watch, follow these steps:

1. **While viewing the default (Infograph Modular) watch face, press and hold the screen.**

   This step enables Force Touch and launches the Face Gallery.

2. **Swipe left or right to select a clock face you like, and then tap the center of the watch face you want to use this time.**

   You can also tap New (+) to select a face.

3. **Tap Edit near the bottom of the screen to personalize the face.**

   You can swipe left or right to see different backgrounds (if available) and twist the Digital Crown to rotate between colors. (In Figure 4-41, you can see I'm choosing the purple background.)

FIGURE 4-40:
The Face Gallery section of the Apple Watch app (on the iPhone).

4. **Flick the screen left and right, and twist the Digital Crown button to customize the face.**

   When you like what you see — after changing the color of the hands or text or adding a second hand, for example — swipe to the left to go to the next customization screen. I cover customizations in greater depth in the later section "Differentiating Between Customizations and Complications."

5. **Twist the Digital Crown button to make your selection again.**

   Repeat this step until you've gone through all the customization screens, perhaps choosing color and extra time information. Typically, the last customization page is for complications.

6. **Tap the highlighted areas that you can add to your watch screen, and then twist the Digital Crown button to select what you're happy with.**

   Repeat the process by tapping the other areas of the face you want to change, which vary by the watch face you choose. You won't have the same options for all watch faces.

FIGURE 4-41: You can customize most watch faces by tapping Edit, and then swiping left or right and twisting the Digital Crown.

7. **Press the Digital Crown button when you're done customizing your watch face.**

   This step confirms that you've finished with your options and are ready to see your customized watch face.

8. **Tap the center of the screen to confirm your changes.**

   After you've set your watch face, you can tap each of the complications — such as weather or stock quotes, as shown in Figure 4-42 — as discussed in further detail in "Differentiating Between Customizations and Complications."

That's it! That wasn't so difficult, right?

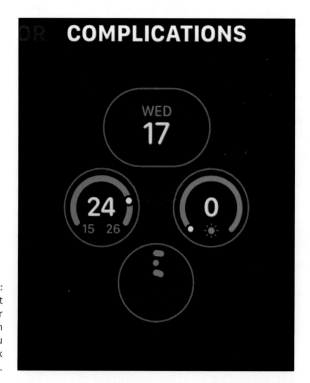

FIGURE 4-42: If you don't like your customization choices, you can tweak them.

Differentiating Between Customizations and Complications

It's important to understand that all the watch faces support various tweaks that can make your Apple Watch feel even more personal. Apple divides these personalized changes into two main categories: customizations and complications.

Customizations (Edits)

Each watch face includes customizations you can perform, such as changing the watch to show additional information (such as specific minute, second, or millisecond detail) and changing the color of the watch hands (perhaps to match your outfit). You can make changes for each face, but what you can change varies by face style. See Figure 4-43 for customization examples.

FIGURE 4-43: Customize the appearance of your watch face by adjusting the amount of information you want to see (left). You can also tweak the color of the hands via the Digital Crown button (right).

Complications

The word *complications* is a horological term that refers to any clock feature that goes beyond the display of hours and minutes.

When you're customizing your watch's face, complication options let you add information to the screen, usually in up to four corners or at the bottom of the screen. This information may include calendar alerts, an alarm, the current moon phase, the weather, the day's sunrise and sunset times, a timer or stopwatch, your current activity, World Clock, and stock quotes. To customize your display, simply touch highlighted parts of the watch face to make your selection. (Be aware that some watch faces offer more room for complications than others.) When you tap the piece of information provided, such as the weather, Apple Watch opens the corresponding app for a deeper dive. Figure 4-44 and Figure 4-45 show a few examples.

FIGURE 4-44:
What can you choose to place on your watch face? A lot, as you see here.

TIP

You can still tell the time on Apple Watch if the battery is low. Your watch automatically goes into Power Reserve mode when your battery drops to a certain percentage, or you can activate this mode manually by opening the Settings app on Apple Watch. You should still be able to see the time for up to 48 hours, according to Apple.

FIGURE 4-45:
An example of
a complication
you can add to
a watch face,
whether you
prefer to do it
on the Apple
Watch itself or
via the Watch
app on iPhone.

Adding Complications

You can add more than 50-odd complications to Apple Watch if they're supported by your specific Apple Watch model and the face you've selected. Table 4-1 lists a handful of them.

Here are a bunch of other complications that may be available, depending on the watch face you've selected: Air Quality · Audiobooks · Battery · Blood Oxygen · Breathe · Calculator · Camera Remote · Cellular · Compass · Cycle Tracking · Date · Earth · ECG · Elevation · Find People · Heart Rate · Home · Mail · Maps ·

Medications · Messages · Music · News · Noise · Now Playing · Phone · Podcasts · Radio · Rain · Reminders · Remote · Shortcuts · Sleep · Solar · Solar System · Stopwatch · Timer · Voice Memos · Walkie-Talkie · Weather Conditions · Wind · Workout

TABLE 4-1   ## A Sampling of Apple Watch Complications

Complication	Function
⏰ **7:00**	**Alarm:** As the name suggests, you can set an alarm, such as that dreaded wakeup call, and see what time it's set for.
🕐 **14:59**	**Timer:** Set one or more timers right on your Apple Watch, and you should see the timer icon count down to zero from your set time (minutes/hours). No more burned cookies!
⏱ **00:00**	**Stopwatch:** Ready, set, go! Right from the watch face, you can start and stop a digital stopwatch. As you likely know, numbers climb up from zero over time and can be reset to start again.
LON 5:09	**World Clock:** Have relatives in another country? Work associates on a different continent? Set a secondary clock on your watch face, and you'll see an abbreviated name for a location (such as LON for London) and the correct time there.
☀ **7:10**	**Sunrise/Sunset:** The name of this complication is also the name of a famous song from *Fiddler on the Roof.* It lets you see when the sun is set to go down and rise the next day.
🌑	**Moon Phase:** Werewolves, rejoice! This complication shows you how much of the moon will be visible that night, such as crescent, half, or full.
MON 9	**Calendar:** See when your next appointment is and what it's about by syncing information from your iPhone's Calendar app. Ideally, it's nothing embarrassing your colleagues that may accidentally see (such as "2 p.m.: Proctologist Appointment")!
72°	**Weather:** Without even having to open an app, your Apple Watch can tell you the weather outside based on what city you're in (or another city). You can choose to see the information in Fahrenheit or Celsius. Ideal for Canadians, eh?
AAPL +.94	**Stocks:** Should you celebrate or bite your nails? The Stocks complication shows you real-time quotes based on the publicly traded companies that matter to you.
◎	**Activity:** Get a quick idea of how active you've been throughout the day. The three rings show your actual level of physical activity and how much of your daily goal you've accomplished. Lazy types might be tempted to throw the watch out, but it won't berate you.
AQI 32 UVI 3.6	**UV Index:** Glance at your local AQI (air quality index) or UVI (ultraviolet index) as complications in the bottom right and bottom left sections of the Solar Dial watch face. With both these indices, the higher the number, the more risk of harm from poor air quality and unprotected sun exposure, respectively.

Accessing Time on Apple Watch

If you've mastered how to select a watch face and customize it, you might want to know how to access the clock on your Apple Watch. After all, it's probably the screen you're going to see most often, right?

Checking the time is quite easy, even if you're in another app. By default, Apple Watch shows you the clock screen when you raise your wrist. Therefore, you needn't do anything if you like this setup. The screen stays on for about four to six seconds, whether you look at it or not, and then goes back to sleep again (presumably to save power).

If you tap the screen or press the Digital Crown button or the side button, however, the screen stays on for about 15 to 17 seconds before fading to black.

Want to see something other than the time when you raise your wrist? No problem. Grab your iPhone (because you need it to change the default setting to another option) and then follow these steps:

1. **In the Apple Watch app on your iPhone, choose My Watch ⇨ Settings ⇨ Display & Brightness.**

    Near the bottom of the screen, you should see a section called **Wake**, as shown in Figure 4-46.

2. **If you like seeing the time when you raise your wrist, do nothing.**

    If not, make some changes by selecting one of the other options, such as seeing the last app you were in.

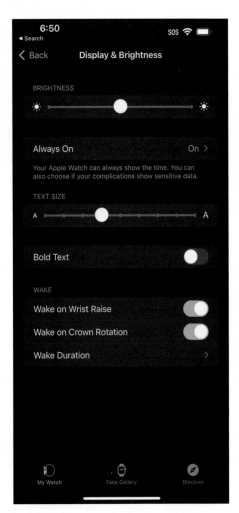

FIGURE 4-46:
The Apple Watch app on iPhone. Here, you can enable and disable various Apple Watch functions.

You can also see the time easily whenever you're in any app (as described in Chapter 3). Simply double-press the Digital Crown button. The watch face immediately appears on your screen. Now double-press the Digital Crown button again, and you return to the last app you were in.

TIP

You have a third way to get the time on your Apple Watch, and you don't even need to look at your wrist at all. Can you guess what it is? Give up? You can ask Siri what time it is. Press and hold the Digital Crown button and ask "What time is it?" or simply say "Time." Alternatively, you can ask "What time is it in [a city around the world]"? Or you can raise your wrist and ask "Hey Siri, what time is it?" See Chapter 7 for more on using Siri to complete tasks with your Apple Watch.

Accessing World Time

Some people like to know what time it is in another part of the globe. You know, in case you want to Skype or FaceTime with someone, and you're not sure whether it's the middle of the night where that person is.

Whether you use it for personal or professional reasons, a world clock is a handy thing to have, and your Apple Watch can help. Introduced in watchOS 8 is support for all 24 time zones at once. The locations around the outer dial represent the different time zones, while the inner dial shows the current time in each location. Tap the globe in the center to see the time zone you're in (which is also noted by the arrow at 6 o'clock).

In "Differentiating Between Customizations and Complications" earlier in this chapter, I discuss adding World Clock as a complication to an existing watch face, but you can also view it on your own.

To use the World Clock app on your Apple Watch, follow these steps:

1. **Press the Digital Crown button to go to the Home screen.**

    Regardless of the app you're in, you should see the screen with all the small icons on it after you press the Digital Crown button. If you don't see what you want at first glance, swipe your finger around to view other bubble-shaped icons.

2. **Tap the World Clock app.**

    The app launches full-screen, and you should see the time in another city. If you haven't added another city yet, Apple Watch might say **No World Clocks**, as it does on an iPhone.

3. **If you have multiple cities selected, you can swipe left or right to navigate among them.**

   The top right section of the screen shows your local time. Then you should see the remote city highlighted by an orange dot on a map, the name of the city, the time there, the time zone, and sunrise and sunset information, as shown in Figure 4-47.

4. **If you don't have any locations installed, or if you'd like to add more (or remove one), open the Settings tab on your iPhone's Apple Watch app, and make your changes in the General section.**

   After you've made your selection, you can exit the app.

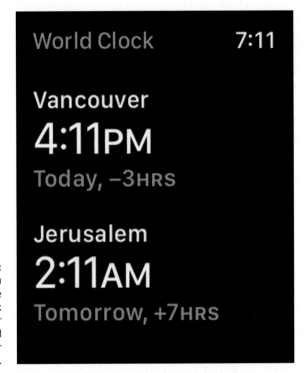

FIGURE 4-47: When you launch the World Clock app, your screen should look similar to this.

You can also ask Siri for this information, such as "What time is it in Warsaw?" or "What time will the sun rise in Tokyo?" See Chapter 7 for more tasks Siri can help you with.

Taking Control: Alarms, Stopwatches, and Timers, Oh My!

Alarms, timers, and stopwatches are great tools to have with you. And because you're wearing Apple Watch on your wrist, you have access to these tools wherever life takes you. I look at these tools, including the new Shortcuts feature, in this section.

Alarms

Apple Watch makes it easy to set an alarm, such as a 7 a.m. wakeup call, which you can view in either analog or digital display. To change between digital and analog, press firmly on the screen and then tap Customize. Swipe left until you see the alarm face you like.

To set up a new alarm on Apple Watch, follow these steps:

1. **Press the Digital Crown button to go to the Home screen.**

2. **Tap the Alarm app to launch it.**

   From this screen, you can view, manage, and edit multiple alarms with your fingertip.

   To set an alarm with Siri, press and hold the Digital Crown button to activate Siri, or simply raise your wrist and say "Hey Siri," followed by "Set an alarm for [date and time]." You can also say "Wake me up at [time of day]" or even "Wake me up in [minutes or hours]."

   Siri confirms the time and shows it to you on the watch's screen, as shown in analog mode on the left side of Figure 4-41 (see "Complications" earlier in this chapter). It's totally fine to have multiple alarms.

3. **To set your alarm time, twist the Digital Crown button to adjust the hours and minutes, and then tap Set.**

   The right image in Figure 4-48 shows an alarm setting.

   Your alarm is set. You can deselect any alarms that you've already set and no longer need. You can also press and hold the Apple Watch screen in the Alarm app to bring up options, including alarm repeats (such as for weekdays). When the alarm goes off, you can tap Snooze or Dismiss.

FIGURE 4-48:
The analog
mode of the
Alarm screen.
Set your
desired time,
using the
Digital Crown
button (left).
It's a cinch to
adjust your
alarm time,
even if you're
half asleep and
fumbling to set
a wake-up call
(right).

TIP

Your iPhone alarms can also be synced with your Apple Watch. Synchronization happens automatically by default, but you can also go into your Apple Watch app on an iPhone and tap My Watch ⇨ Settings to disable this feature.

Stopwatches

You don't need to be an Olympic runner to appreciate a stopwatch, which measures the ascending passage of time. Whatever the reason you'd like to know how much time has elapsed, the Stopwatch app is what you need — and it's fully customizable, too.

That is, the Stopwatch app on Apple Watch lets you see information in a digital, analog, hybrid view, or even in a graph that shows a real-time average of your lap times. Figure 4-49 gives you a look at the hybrid view.

To use the Stopwatch app on Apple Watch, follow these steps:

1. **Press the Digital Crown button to go to the Home screen.**

2. **Tap the Stopwatch app.**

   The Stopwatch app launches. You can also raise your wrist and then say "Hey Siri, Stopwatch," or press and hold the Digital Crown button to initiate Siri.

FIGURE 4-49:
You can
choose a
hybrid of
analog and
digital, as
shown here.

3. **Tap the green Start button in the bottom-left section of the screen, as shown in Figure 4-50, to start the stopwatch.**

   Whether you're in analog, digital, or hybrid view (which you can change in the Settings area of the iPhone's Apple Watch app, as discussed in Chapter 11), you should see the time scroll by.

4. **Press the red Stop button in the bottom-right corner of the screen to stop the stopwatch.**

   Don't worry if you accidentally close the app without taking note of your time; it'll still be there when you open the app again.

5. **Tap the Lap button in the bottom-right corner of the app if you want to see graphed averages of lap times.**

   Joggers and runners might appreciate this added historical information, shown in Figure 4-51.

FIGURE 4-50:
The Stopwatch app in action.

FIGURE 4-51:
The Stopwatch app offers a historical/graphical look of your lap times.

Timers

The Timer app lets you keep track of events you want to . . . well, time. Whereas the Stopwatch app measures the ascending passage of time, the Timer app offers the descending passage of time from a preset starting point, such as 45 minutes.

As with the other time-related apps, you can choose to read a digital or analog countdown.

To use the Timer app on your Apple Watch, follow these steps:

1. **Press the Digital Crown button to go to the Home screen.**

2. **Tap the Timer app.**

   The Timer app launches. Or you can say "Hey Siri, Timer" or press and hold the Digital Crown button to activate Siri and then say "Timer."

3. **Twist the Digital Crown button to select the starting time.**

   You start with the hour setting. Press the Digital Crown button to go to the minute setting and then press the Digital Crown button again when you're done. Review what you selected for a start time, such as 30 minutes or an hour and 15 minutes.

4. **Tap the green Start button in the bottom-left section of the screen to initiate the timer.**

   As the timer runs, an orange line moves around the dial in clockwise fashion to give you a visual sense of how much time has passed (and how much is left). Figure 4-52 shows the Timer app in all its glory.

5. **When you're done, press the Reset button in the bottom-right corner of the Apple Watch screen.**

   Pressing Reset rolls back the timer to all zeros, making it ready for the next time. If you no longer need the timer, press the Digital Crown button.

Beginning with watchOS 8, you can use Siri to give timers a specific label (such as "Oven Timer" or "Laundry Timer").

FIGURE 4-52:
The Timer app
offers different
views and even
the ability to
customize
(and name)
different
timers.

Shortcuts

If you haven't yet taken advantage of Shortcuts on your iPhone or iPad, a feature first introduced in iOS 12, it lets you automate specific tasks — or sequences of tasks — you can trigger with a single tap, Siri voice command, and now via Apple Watch. You might build a "Dinner Prep" shortcut that turns on the lights in the kitchen or plays your favorite music playlist on your kitchen smart speaker.

The Shortcuts app lets you initiate a shortcut with a tap on the screen. Or you can tell Siri to initiate it. Or you could add a shortcut to a watch face as a complication.

The Apple Watch app offers a couple of default shortcuts to get you started. Open the Shortcuts app from the Home screen, and tap a shortcut to run it (see Figure 4-53).

To add a shortcut complication to a watch face, touch and hold the watch face, and tap Edit. Swipe left to go to the Complications screen, and then tap the complication. Then scroll to Shortcuts and choose a shortcut.

If you want to branch out from the default shortcuts, you can add a shortcut you have stored on your iPhone. Open the Shortcuts app on your iPhone, tap More in the shortcut's top-right corner, and then tap Show on Apple Watch.

REMEMBER

Not all iPhone shortcuts are compatible with Apple Watch.

IN THIS CHAPTER

» Accepting and placing calls on your Apple Watch

» Handing off calls to your iPhone

» Receiving and sending messages, including the Sketch feature

» Sending emojis, Animojis, and Memoji stickers

» Sharing voice clips, taps, heartbeats, and more

» Using the Walkie-Talkie feature

» Receiving and managing emails

Chapter **5**

Keep in Touch: Using Apple Watch for Calls, Texts, and More

Apple Watch isn't used just for information such as the time, the weather, stock quotes, and sports scores. It's also ideal for keeping in touch with those who matter.

In other words, your smartwatch isn't just about knowledge; it's also about communication. That's precisely what this chapter is all about. On one hand (if you can pardon the pun), you've got the same familiar ways to connect that you have on your iPhone. Specifically, your Apple Watch can place and accept calls as well as send messages and emails. Although Apple Watch was designed for quick interactions, it supports many of the same chatting features as your smartphone.

In fact, if you own one of the Apple Watch models with GPS + Cellular — such as Apple Watch Series 8, Apple Watch SE, or the large Apple Watch Ultra — and have set up a wireless plan with your phone provider, you don't even need your iPhone nearby to place or accept a call or text message! But you can also use your watch to reach out to others in new and unique ways, such as sending someone a tap or your heartbeat, which they feel on their own wrist (if they're wearing an Apple Watch). You can also send a finger-drawn sketch to someone special.

Note: Apple Watch Ultra is the only Apple Watch that automatically includes GPS + Cellular in every model, but it's up to you if you want to pay for a monthly cell phone service.

Accepting and Placing a Call on Apple Watch

If you want to be like Dick Tracy and take calls on your wrist, Apple Watch lets you do just that. Or you can make a call by pressing the Digital Crown button and asking Siri to call someone.

You have two ways to make calls, depending on which model of Apple Watch you have:

>> **With cellular connectivity:** Whether you initiate the call or accept it, as long as you have a cellular plan on a supported Apple Watch (GPS + Cellular model), you can chat through your smartwatch's microphone and hear the other person through the small speaker. When your watch connects to a cellular network, four green dots appear in the watch's Control Center.

>> **Without cellular connectivity:** If you don't have an Apple Watch that supports cellular connectivity, you can still use your Apple Watch to chat with others. You handle this task via Bluetooth technology with your nearby iPhone — up to a few dozen feet away — or even farther than that over Wi-Fi. As long as your iPhone is connected to the same wireless network, such as at home or at the office, your watch rings at the same time as your phone, and you can choose which one to answer (or not).

Incoming calls

This part is the easy one.

If a call comes in to your phone number, your Apple Watch rings just like your iPhone — unless you choose to disable that feature in the Apple Watch app on your iPhone, which you can find out how to do in Chapter 11. You'll also feel a slight haptic vibration on your wrist when a call comes in, and this setting is also something you can change if you like. Assuming you didn't mess with the default settings, you hear your ringtone emanate through your watch's speaker and see a screen pop up with the name of the person calling (or just a phone number if that person isn't in your Contacts list). Tap the big green icon in the bottom-right section of your watch's screen to answer the call.

If you don't want to answer the call, tap the big red (Hang Up) icon in the bottom-left section of the screen. Alternatively, you can swipe up from the bottom of your watch to send a preset message back to the person who's trying to reach you, such as "Call you later," "In a meeting," or "I'm driving." Just make sure that the call is coming from a mobile phone; otherwise, the caller may not see it.

Figure 5-1 shows a couple of preset messages you can send if you're unable (or unwilling) to speak.

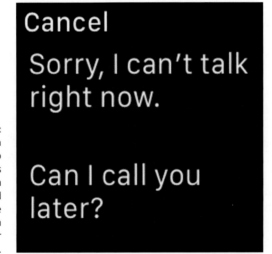

FIGURE 5-1: You can respond to incoming calls by sending a customized reply in the Apple Watch app on your iPhone.

You can customize these preset replies by tapping Settings ⇨ Messages on your iPhone; see Chapter 11 for more information. But as infomercial guru Ron Popeil once famously said, "But wait, there's more!":

>> **Transfer it.** You can transfer your call to your iPhone, a Bluetooth headset, or a car's speakerphone. See "Handing Off a Call to Your iPhone or via Bluetooth" later in this chapter.

>> **Stop your wrist from ringing.** Perhaps you're in a crowded elevator, and you're getting dirty looks from someone. You can silence an incoming call simply by covering Apple Watch with your other hand. (You could try to cover it with the same hand the watch is on, but that may prove to be a tad difficult.)

>> **Mute the ring.** The Phone app screen on Apple Watch lets you mute your microphone by tapping the microphone with a slash through it (in case you don't want other participants in a conference call to hear you sneeze!). At the top of the watch screen, you can increase or decrease the volume coming through the watch's speaker.

Outgoing calls

Although it's certainly not difficult, placing outgoing calls through your Apple Watch requires a little more work, and that's assuming that you actually *want* to make a call through your wrist.

A few considerations: It may not be too comfortable to hold up your wrist for an extended period; the quality of the call won't be as good as on a phone; your conversations may be overheard because the watch uses a speakerphone (unless you're wearing a Bluetooth headset); and you might look a little silly, too.

The Apple website also suggests that you may not want to talk long on Apple Watch anyway: "Use the built-in speaker and microphone for quick chats, or seamlessly transfer calls to your iPhone for longer conversations." With that in mind, you can place a call on your Apple Watch in a few ways:

>> **First approach:** Press the Digital Crown button to go to your Home screen; then tap the Phone icon. You will see: Favorites, Recents, Contacts, and Keypad. Go ahead and dial the number in the Keypad section, or pull up someone from one of the other sections (see Figure 5-2). Swipe down or twist the Digital Crown button to access one more option: Voicemail.

Figure 5-3 shows you the other options that appear when you pull up a contact in the Phone app. When you're in the Contacts part of the Phone app, you can see the icons for placing a call, texting (via Short Message Service [SMS]) someone, or sending an email. Below that section is Notes, if you've ever filled out that section on your iPhone. Yes, your Contacts info is synced with the Contacts app on your iPhone!

>> **Second approach:** Press the side button to bring up Dock, and twist the Digital Crown button or swipe up and down to get to the Phone section (if you recently used the watch to make a call or set the Phone app as one of your favorite Dock apps). Tap the Call icon for a given contact, as shown in Figure 5-4.

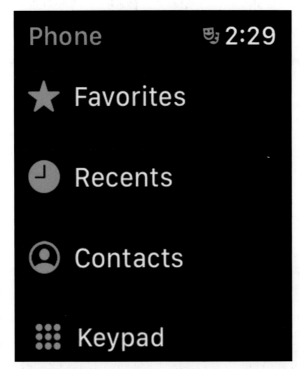

FIGURE 5-2:
When you
open the
Phone app,
you see the
main screen.

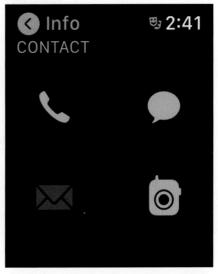

FIGURE 5-3:
Pulling up a
contact in the
Phone app
reveals icons
for calling or
texting (left),
as well as for
emailing and
viewing notes
(right).

FIGURE 5-4:
Access the
Phone section
of Apple Watch
by pressing the
side button.

>> **Third approach:** Lift your wrist and say "Hey Siri" into your watch, followed by "Call [person's name]" (if that person is in your phone's Contacts list) or "Dial [phone number]." Or you can press and hold the Digital Crown button to activate Siri. Or you can say "Hey Siri, make a call" and wait for it to ask you to name someone or provide a number, as shown in Figure 5-5.

Outgoing calls with FaceTime audio

You can't make video calls from your wrist on Apple Watch — er, yet — but you can use FaceTime to place an audio call, if you like. You may be in a Wi-Fi hotspot and prefer to call over FaceTime (which uses data) over a cellular connection. Or the person you're calling may be using an iPad, in which case FaceTime is the only way to call them (see Figure 5-6). I whited out the phone numbers for this contact for privacy reasons, but you get the gist.

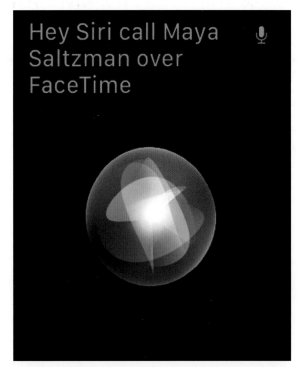

FIGURE 5-5:
The fastest way to make a call? Use Siri to dial a contact or phone number (actual phone number blocked out).

FIGURE 5-6:
It's super-easy (and, frankly, fun) to place a FaceTime audio call from your wrist. If the person you're calling has FaceTime, it will appear as an option.

To place a FaceTime call by using the Phone app on your Apple Watch, follow these steps:

1. **Open the Phone app on your Apple Watch.**

   Follow the directions in "Outgoing calls" earlier in this chapter.

2. **Tap Contacts, and tap the contact you want to call.**

3. **Tap the Phone icon.**

   You may see an option for FaceTime Audio (if your contact has FaceTime).

4. **Tap the FaceTime Audio option.**

Alternatively, you can summon Siri and ask her to place a FaceTime audio call with someone too! Siri is the easiest method. To recap, with examples, raise your wrist or say "Siri," followed by something like

>> "Call Mom."

>> "Dial 800 555 1212."

>> "Call Pete FaceTime audio."

TIP

If you ever want to turn the Apple Watch screen off or silence any sounds coming from it, simply cup your hand over the face. The screen should go dark and silent.

Not too confusing, right? You can always revert to smoke signals or Morse code if all this technology is making your head spin!

Handing Off a Call to Your iPhone or via Bluetooth

You probably don't want to talk for long periods through your Apple Watch, if only because long calls eat up the battery. The solution is wirelessly handing off the call to your iPhone.

By the way, the Handoff feature is available in other apps too, not just for phone calls. It's ideal when you want to transfer what you're doing to a nearby iOS device (iPhone, iPad, or iPod touch) or Mac. Handoff is part of Apple's Continuity feature over Wi-Fi and includes such apps as Calendar, Reminders, Messages, Mail, Contacts, and Maps.

Handoff should already be enabled on your Apple Watch, but if it isn't for whatever reason, Figure 5-7 shows how you can enable it in the Apple Watch app on your iPhone. Tap My Watch ➪ Settings ➪ General, and flick the Enable Handoff slider button to green.

If an Apple Watch app can be handed off to your iPhone, you should see the Handoff option in the watch app. Select that option and tap the icon in the bottom-left corner of the iPhone screen to complete the handoff to your phone.

Apple says this about Handoff: "When this is on, your iPhone will pick up where you left off with apps on your Apple Watch. Apps that support this feature appear on the lower-left corner of your iPhone lock screen."

To use Bluetooth to hand off a call, when a call comes in, slide up from the bottom of the watch screen to see a list of options, including one that lets you hand the call to another Bluetooth-enabled device: the iPhone itself, a hands-free Bluetooth headset, or perhaps a Bluetooth-enabled stereo in your vehicle.

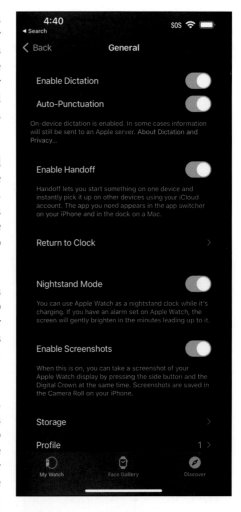

FIGURE 5-7:
You can tweak several options in the Apple Watch app on your iPhone, including transferring activities to other Apple products.

Making Apple Watch Calls over Wi-Fi

If your carrier supports Wi-Fi calling (as many carriers do), you can place and receive calls over Wi-Fi rather than a cellular network. Wi-Fi is especially handy if you don't own an Apple Watch with GPS + Cellular, because you can use your Apple Watch to make calls even if your iPhone isn't nearby. As long as you're in a Wi-Fi network, you're good to go.

To get going, follow these steps:

1. **On your iPhone, tap Settings ⇨ Phone ⇨ Wi-Fi Calling; then turn on both Wi-Fi Calling and Add Wi-Fi Calling for Other Devices.**

    See Figure 5-8. If you can turn this feature on, your carrier supports it.

2. **Enable Wi-Fi calling on your Apple Watch by opening the Apple Watch app on your iPhone, tapping My Watch ⇨ Phone, and then turning on Wi-Fi Calls.**

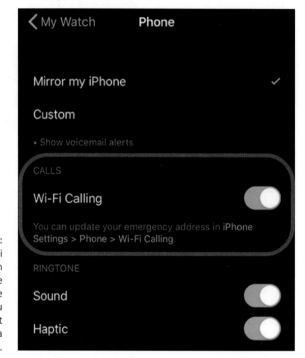

FIGURE 5-8: Enabling Wi-Fi calling on both your iPhone and Apple Watch lets you place or accept calls over a Wi-Fi network.

TIP

I recommend that you make emergency calls over a cellular connection rather than Wi-Fi because your location information is more accurate. To err on the side of caution, make sure that your emergency address is up to date: On your iPhone, tap Settings ⇨ Phone ⇨ Wi-Fi Calling ⇨ Update Emergency Address.

Since the fall of 2019, Apple Watch has been able to make international calls to emergency services regardless of where the device was purchased or whether the cellular plan is activated.

Ideal for world travelers, international emergency calling also works with fall detection (if that feature is enabled) to place an emergency call if Apple Watch senses that the user has taken a hard fall and remains motionless for about a minute (see Chapter 8).

Receiving and Sending Messages

Many millions of iPhone users send messages from one iPhone to another device, whether those messages go through the standard SMS service or Apple's own iMessage service. The distinction may be a bit confusing because the app itself is called Messages, but iMessage is part of that service, along with regular texting. The main advantage of iMessage over texting is that it's free for you and anyone else over Wi-Fi, and it's unlimited, so you can type and upload media as much as you want.

Apple Watch also houses a Messages app — supporting both iMessage and text messages — that you can use to chat with someone while on the go.

Although you can type on the larger Apple Watch Series 7 or Series 8 only by pulling up the QWERTY keyboard, any Apple Watch lets you send messages through voice dictation (or send the audio clip instead), tap preset responses based on the context of messages you've received, use the Scribble feature with your fingertip (see below), send animated GIFs and emojis (cute and customizable icons), and even share your location on a map, which can be quite handy (as I cover in Chapter 6).

In fact, with the introduction of watchOS 8, you can even combine the use of dictation, and emojis — all within the same message!

Receiving and responding to messages

You can keep your iPhone tucked away yet still correspond with important people in your life. When a message comes in (via iMessage or SMS), your Apple Watch vibrates on your wrist and dings to let you know that you have a new message waiting to be read. (Again, you can disable tactile feedback and sound in the Settings area of the Apple Watch app on your iPhone, as discussed in Chapter 11.)

Acknowledging a message

To receive, reply, and initiate a message to someone via your Apple Watch, follow these steps:

1. **If you feel a pulse and hear a tone, raise your wrist to see the message.**

   You should see a screen (see Figure 5-9) showing the sender's name, what they wrote, and perhaps an integrated image. You can scroll up and down by twisting the Digital Crown button if the message is longer than what's on the screen.

2. **To dismiss the message, lower your wrist or tap Dismiss at the bottom of the message.**

   Do either thing if you don't want to reply — or at least not right now. You can exit the Messages app and return later.

3. **If you want to reply to the message, tap the small reply field at the bottom of the screen.**

   Now you should see several "suggested" options on how to reply (see Figure 5-10). Apple Watch suggests some preset words to reply with, based on the context of the conversation, along with some preset responses you can choose from ("Not sure," "Can't talk now," and "Talk later?"). Twist the Digital Crown button to see all the responses; then tap one you like. To create a custom response in the Apple Watch app on your iPhone, see Chapter 11.

Replying to a message by typing

On Apple Watch Series 7, Apple Watch Series 8, and Apple Watch Ultra, tap the Create Message field, then use the QWERTY and keyboard that pops up. Tap characters to enter them one at a time, or use the QuickPath feature to slide from one letter to the next without lifting your finger. To end a word, lift your finger.

Replying to a message by 'scribbling'

On Apple Watch Series 7, Apple Watch Series 8, and Apple Watch Ultra, swipe up from the bottom of the screen to change from keyboard to Scribble.

If you aren't on Apple Watch Series 7, Apple Watch Series 8, or Apple Watch Ultra, you can tap the Scribble options, which lets you use your finger to handwrite your message, letter by letter. To edit your message, turn the Digital Crown to move the cursor into position, then make your edit.

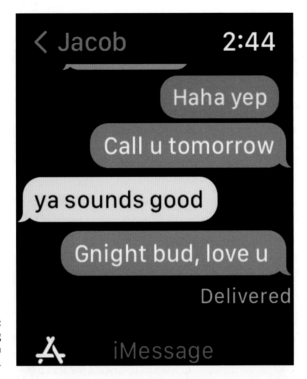

FIGURE 5-9:
An incoming message on Apple Watch.

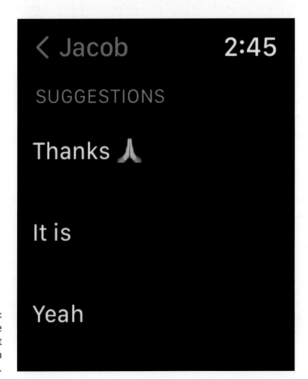

FIGURE 5-10:
You can use several preset replies to a message.

If you've set up your Apple Watch to use more than one language, you can choose a different language when using Scribble by simply swiping up from the bottom of the screen and choosing another language.

Replying to a message by speaking

You can respond to text messages by speaking your reply. Here's how:

1. **Tap the microphone icon to verbally say your reply, and have it transcribed into text or sent as an audio clip.**

   Speak clearly to your wrist, and you should see the words typed as you say them.

   Starting with watchOS 8, you can preview what you said to make sure that everything is correct. If not, you can tap to edit the words before you send, or tap Cancel in the top-left section of the screen and then speak again (perhaps slower and more clearly!).

2. **Tap Send in the top-right section of the screen.**

   If you don't see an option to send your message as a voice clip, tap Settings ⇨ Messages in the Apple Watch app on your iPhone and select whether you want your watch to transcribe your voice as text, and send it as a voice recording.

See "Responding with an audio clip" later in this section for more on sending an audio clip.

Replying to a message via your iPhone keyboard

When you start composing a message and your paired iPhone is nearby, a notification appears on the iPhone, offering to let you enter text using the iOS keyboard. Tap the notification, then type the text on your iPhone.

Responding with emojis or Memoji stickers

To reply to a message with an emoji or animated Memoji stickers (which you can also create), follow these steps.

The first step is for emojis only!

1. **Tap the speech window as if you were going to type a reply with words.**

2. **Now tap the smiley face on the Apple Watch screen to launch the emoji selection list.**

   Figure 5-11 shows some sample emojis, which allow you to express yourself in a more playful way.

3. **Twist the Digital Crown button to select the emoji that conveys your message or feelings, whether it's a smile, a silly tongue hanging out, a sad face, a heart, or something else.**

The small green bar in the top-right section of the screen shows you where your list of options starts and ends. Leave the emoji on the screen for a moment to see how it'll animate when someone else receives it.

4. **When you find an emoji that fits the bill — maybe a thumbs-up or thumbs-down — tap Send in the top-right section of the screen.**

Figure 5-12 shows more examples. If you decide against sending the emoji, tap Cancel in the top-left section of the watch screen.

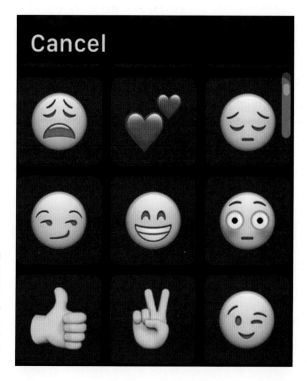

FIGURE 5-11:
Tap the emoji icon to send a playful smiley face or another emoji instead of (or in addition to) your words.

Note: For animated Memoji stickers, the process is similar but when you want to reply to a message, don't tap inside the reply window first, which is what you would do for an emoji. Instead, tap the little symbol in the lower left of the screen that looks like a letter "A" with three sticks, and then tap the icon that looks like faces (including someone with heart eyes). Now you can select a Memoji to send!

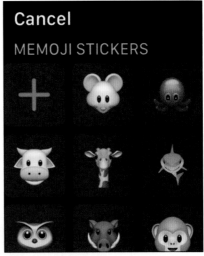

FIGURE 5-12:
Select a regular emoji (left) or an animated Memoji (right).

Replying with a Sketch

If you don't feel like dictating your reply to a message, you can use the Sketch feature to handwrite your words — letter by letter — using your fingertip.

To use the Sketch feature, follow these steps:

1. **Tap an incoming message.**

   Several options appear.

2. **Tap the option that looks like a capital letter "A" made out of three sticks, (left image in Figure 5-13).**

3. **Tap the icon of a heart with two fingers touching it.**

4. **Use your fingertip to draw or write something out.**

5. **When you're done, let go, and it'll send your creation.**

You can draw a smiley face, a star, a heart, a flower, a sun, a fish, written-out words, or anything else you can think of. See Figure 5-13 for an example. Because your friend is wearing an Apple Watch, that person sees the drawing appear on their wrist just as you drew it. They know it's from you because your name is in the top-left corner (which is the same for a tap and a heartbeat).

When sending a sketch, you can tap the small circle at the top right of the Apple Watch screen to change colors.

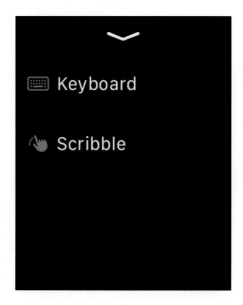

FIGURE 5-13:
Select Sketches
to write
something with
your fingertip
or draw a
picture.

Responding with an audio clip

Unless you want to handwrite your response to a message by using Sketch — which takes some time to do — you're likely to dictate a reply and tap Send. If you're not in a place that prevents talking (such as an important board meeting), you can review your words before you send a reply.

But wait, there's another option! As I touch on in earlier sections, you can speak your reply to a message and send it as a voice clip. This feature is particularly useful for those times when you want the recipient to hear your voice, like a kind of voicemail message.

Here's how to do it:

1. **Read the message someone sent you and then, to reply, tap the symbol that looks like the letter "A" in the bottom right corner of the screen.**

2. **Tap the icon that looks like a bunch of vertical "speech" lines, and start talking.**

   When you're done, tap Done. (Easy, huh?)

Now the screen says **Send As Audio** or (if you opted for both) **Send As Audio** or **Send As Text.** Figure 5-14 shows what your message looks like before you send it, as well as what your recipient sees.

FIGURE 5-14: How a message appears as a voice recording or transcribed text (left and middle).

Responding with a GIF

Whether you call GIFs "jiffs" or "giffs" (with a hard "g") — both pronunciations are acceptable — these trendy, short animated clips are fun to share in iPhone messages or post to social media. Now you can include them in your Apple Watch messages too.

To include a GIF in your message, follow these steps:

1. **Tap the message you want to reply to and tap the icon in the lower left of the screen that looks like a letter "A" made out of sticks.**

   It's okay to do this after you've already started typing out your message but haven't hit Send yet.

2. **Now tap the icon that's red (top-right) and you'll see some gifs populating the screen.**

3. **Tap one you like, swipe down to find more (twist the Digital Crown) or use the Search feature to write a keyword (like "birthday") to find a relevant gif to send.**

   Apple Watch shows some related GIFs that you can use. See Figure 5-15 for an example of what may pop up.

4. **Select the animated gif and send it to the recipient.**

Sending a message

You can send a new message via Apple Watch in two ways: by using the Messages app or by using Siri.

If you have an Apple Watch with cellular support, you don't need your iPhone nearby; if you don't have cellular support, you do!

To send a message from your Apple Watch by using the Messages app, follow these steps:

1. **Press the Digital Crown button to go to the Home screen.**

2. **Tap the Messages app.**

   The Messages app launches.

3. **To send a new message, press and hold the screen until the words New Message pop up (see Figure 5-16); then tap them.**

   Now you're ready to start a new message to someone.

4. **Tap Add Contact to select the recipient.**

5. **Tap Create Message, which lets you dictate your message (also shown in Figure 5-16).**

   Again, you have an option for Apple Watch to transcribe your words as text, or you can send your message as a voice clip.

FIGURE 5-15:
You can use GIFs — animated images — to further express yourself when messaging back and forth with friends and family members over Apple Watch.

To send a message on your Apple Watch by using Siri, follow these steps:

1. **Lift your wrist and say "Hey Siri" into your watch, followed by "Message [person's name or number]."**

   You can also press and hold the Digital Crown button to activate Siri. Figure 5-17 shows an example. The Messages app opens, and you should see the name of the person to whom you want to send a message.

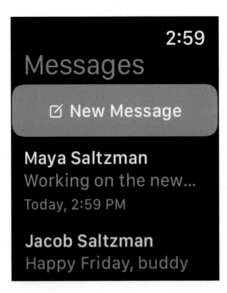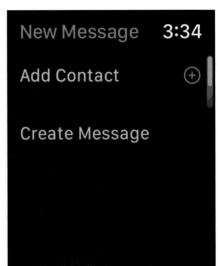

FIGURE 5-16:
When you tap New Message (left) and choose a contact to whom to send a message, you can tap Create Message to compose the message (right).

2. Dictate your message by tapping the microphone icon.

3. Tap Send when you're done recording your message.

   After a few moments, the person you're messaging receives the text or audio clip.

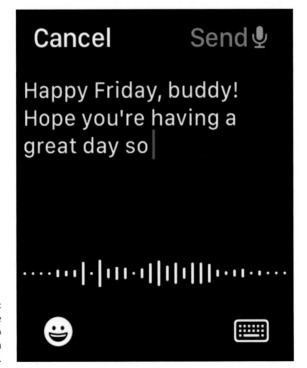

FIGURE 5-17:
Use your voice to reply to messages on Apple Watch.

MESSAGE MIRRORING ON iPhone

Whenever you send a message over Apple Watch, it's mirrored with your iPhone unless you want to turn that feature off. As shown in the figure, you can open your Messages app on your iPhone and see the same conversation with someone as on your Apple Watch.

When the devices are synced, your Apple Watch and iPhone share messages, so you can start on your watch and continue on your phone, if you like. Gee, I hope that Jacob likes my silly animated hamburger. (The eyes jiggle.)

REMEMBER

A super-fast way to send a message to someone through Apple Watch is to raise your wrist and say, "Hey Siri, message [person's name]," followed by the message itself. Siri shows you the message before you send it. See Chapter 7 for more ways to use Siri to help you perform tasks with your Apple Watch.

You can also start a message on Apple Watch and continue it on your iPhone. As you can with calls and emails, you can easily transfer messages to your iPhone, where you can pick up right where you left off. Apple Watch is meant for quick interactions, not lengthy ones. See "Handing Off a Call to Your iPhone or via Bluetooth" earlier in this chapter to find out more about handing off calls to your iPhone from your Apple Watch.

Creating custom replies

If you find it convenient to select one of the many Smart Replies to someone through Apple Watch — such as "Call you later" or "Thanks!" — keep in mind that you can create your own custom replies. You'll need your iPhone.

To customize a reply, follow these steps:

1. **Open the Apple Watch app on your iPhone, and tap the My Watch tab.**

2. **Tap Messages ⇨ Default Replies, and select a default reply to change it.**

   Or tap Add Reply at the bottom of the screen to create your own.

3. **To remove a default reply or change the order, tap Edit.**

4. **To add a custom reply, tap "Add reply . . ."**

   That's it!

Sharing your location from Apple Watch

Suppose that you're chatting with a friend you're supposed to meet up with, but they can't seem to find you. Did you know that Apple Watch lets you share your location with someone instantly? As shown in Figure 5-18, the task is as easy as tapping a contact you're having a conversation with and then scrolling down all the way to the bottom of the screen, and you'll see Send Location (under More). Tap that to share where you are, via Apple Maps.

But wait, there's more! If you press and hold on some blue text in the Messages app on Apple Watch, you'll see some options pop up (see Figure 5-19), including a "thumbs up," "heart" (love), "Haha." Apple calls this "Tapback." You also have the option to copy the text and quickly share with someone else in Messages.

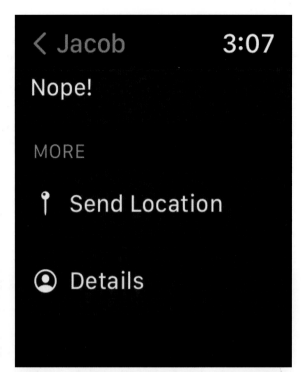

FIGURE 5-18:
Share your geographical location instantly via Apple Maps by scrolling down to the bottom of a conversation with someone and selecting Send Location.

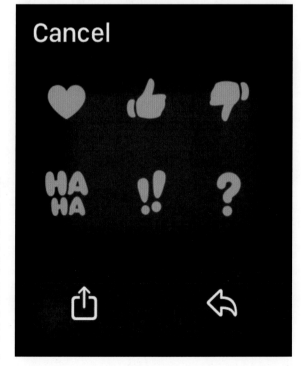

FIGURE 5-19:
Press and hold on a conversation in Messages to send a quick emoji, like a "thumbs up," or the option to forward the note onto someone else in your Contacts.

Sending Taps, Kisses, Heartbeats, and More

Just like your iPhone, Apple Watch lets you communicate via phone and text, but it can also do some things that your smartphone can't do. Collectively, these actions are *Digital Touch* features, available exclusively on Apple Watch.

Call these features wrist-to-wrist communication.

Digital Touch allows Apple Watch wearers to connect with other Apple Watch wearers in fun, unique, and spontaneous ways. Specifically, Digital Touch offers six options, all shown in Figure 5-20.

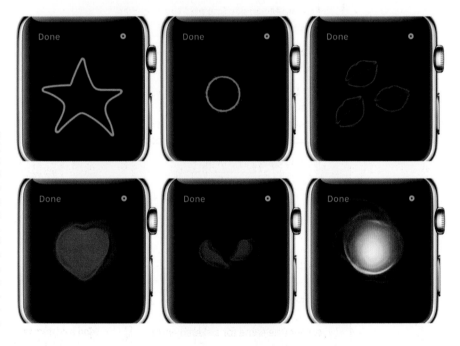

FIGURE 5-20: Use Digital Touch options to communicate from one Apple Watch to another. Clockwise starting at top left: Sketch, Tap, Kisses, Show Anger, Heartbreak, and Heartbeat.

We already covered Sketch, so let's look at the other options:

» **Tap:** Send gentle (and even customizable) taps to someone to let that person know you're thinking about them.

» **Kisses:** Send your sweetheart some virtual kisses directly to their Apple Watch by tapping the screen.

» **Heartbreak:** Want to tell your lover you've been hurt? Your Apple Watch can do this for you with an animated broken heart.

>> **Heartbeat:** Your built-in heart-rate monitor information is captured and sent to someone special so that person can feel it on their wrist.

Tap

Another unique way to use Digital Touch on Apple Watch is to send a tap to someone. As with the other Digital Touch features, that person also needs to have an Apple Watch. A tap is similar to a tactile version of a Facebook "poke" — a kind of "Hey, I've been thinking about you" type of notification.

To send a tap from your Apple Watch, follow these steps:

1. **Tap the Messages app and then select someone's name.**

2. **Select the Digital Touch icon (a heart with two fingers inside).**

3. **Tap the screen one or more times, such as making two quick taps in the top-left corner, and then press the bottom-right section of the screen.**

   Tap the small circle in the top-right corner to change the color. Stop tapping to send.

   When you complete those taps, the person you're tapping sees and feels the taps on their Apple Watch — visually presented as small and large rings that disappear (as shown in Figure 5-21). The recipient knows the taps are from you because your name is in the top-left corner (the same for a sketch and a heartbeat).

FIGURE 5-21:
Send a custom tap to someone else who has an Apple Watch, and that person should see and feel it on their own wrist.

The recipient can "play" the pattern again by tapping the top-right corner of their screen. If and when the person you're tapping with replies with a Digital Touch, they can choose to reply with something else, such as a sketch or a heartbeat instead of another tap.

Heartbeat

Sending your heartbeat is another way to reach out and flirt with someone — from your Apple Watch to another person's Apple Watch. Is that romantic or what!?

To send a heartbeat from your Apple Watch, follow these steps:

1. **Tap the Messages app and then select someone's name.**

2. **Select the Digital Touch icon.**

3. **Press two fingers on the screen at the same time.**

    Apple Watch's heart-rate monitor (which is below the watch face) immediately starts calculating your heart rate.

    After a few seconds, your heartbeat — as shown in Figure 5-22 — is sent to (and felt by) your significant other or friend. They know it's from you because your name is in the top-left corner (the same for a sketch and a tap).

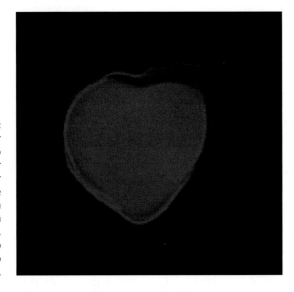

FIGURE 5-22: Send your heartbeat to a friend's or your better half's Apple Watch, which they'll see as a beating heart. You need two fingers to do so.

Okay, sending a heartbeat might sound a little gimmicky, but it might just brighten your friend's or loved one's day if they're having a rotten one. Just don't send it to the wrong person, such as your boss or your best friend's spouse — in case someone gets the wrong idea.

A few more Digital Touch options

Didn't see what you wanted to send your friends and loved ones in the previous sections? Try the following options:

>> **Send a kiss.** Tell someone you're thinking of them by sending kisses. The process is similar to sending a heartbeat. Follow the instructions in "Sending a message" to select a person to message with, and tap the Digital Touch icon. To send kisses, tap two fingers on the screen one or more times. Stop tapping to send.

>> **Break a heart.** Time to channel your inner Taylor Swift and tell someone they've broken your heart? (Yes, I went there.) Follow the instructions in the "Sending a message" section to select a person to message with, tap the Digital Touch icon, and then place two fingers on the screen until you see and feel your heartbeat. Drag down to send.

>> **Show anger.** Don't know how to communicate your fury? Apple Watch to the rescue. Follow the instructions in "Sending a message" to select a person to message with, tap the Digital Touch icon, and then press the screen with one finger until you see a flame. Lift to send.

Enabling and Using the Walkie-Talkie Feature

Walkie-Talkie is a fun way for Apple Watch wearers to communicate. As you might expect, this feature lets you chat with someone wrist to wrist by using your voice (see Figure 5-23).

REMEMBER

The Walkie-Talkie app isn't available in all countries or regions. But no, you don't need to be near the person you're talking to, as you do with old-school walkie-talkies (like those in the *Stranger Things* TV series)!

To get going, you have to fulfill two requirements:

>> Both you and your friend can use any version of Apple Watch, but both watches need the watchOS 5 operating system or later.

>> You both need to set up the FaceTime app on your iPhones and be able to make and receive FaceTime audio calls. See "Outgoing calls with FaceTime audio" earlier in this chapter.

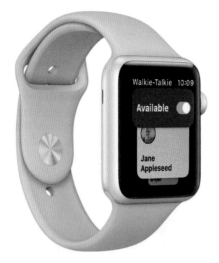

FIGURE 5-23:
10-4, good buddy? You can use the Walkie-Talkie feature to chat with someone via their Apple Watch.

Adding and removing Walkie-Talkie friends

To add a friend, follow these steps:

1. **Open the Walkie-Talkie app on your Apple Watch.**

   The icon is yellow and black.

2. **Press the yellow + sign and choose a contact.**

   Wait for your friend to accept the invitation. The contact card stays gray and is labeled **Invited** until your friend accepts.

3. **After the other person accepts, their contact card turns yellow, and you and your friend can talk instantly.**

   Figure 5-24 shows what this process looks like.

To remove a friend, open the Walkie-Talkie app, swipe left on the friend's name, and tap the red *X*. Alternatively, you can open the Apple Watch app on your iPhone, tap Walkie-Talkie ⇨ Edit, tap the red minus sign (–), and then tap Remove to confirm.

Having a Walkie-Talkie conversation

Now for the fun stuff. Follow these steps to talk:

1. **Open the Walkie-Talkie app on your Apple Watch.**

2. **Tap a friend's contact.**

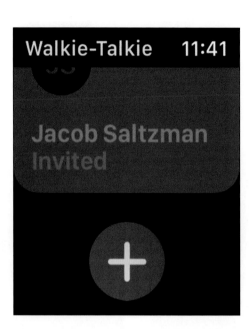

FIGURE 5-24:
Send an invite
to someone
over Walkie-
Talkie (left),
and your
friend's wrist
displays your
request (right).

3. **Touch and hold the Talk button, and then say something.**

   You might see the word **Connecting** on the screen while the watches are attempting to connect wirelessly. After the connection is made, your friend can hear your voice and talk with you instantly.

4. **When you're done talking, let go of the Talk button.**

   Your friend instantly hears what you said.

To change the volume, turn the Digital Crown button.

Need a little peace and quiet? To turn Walkie-Talkie off, open the app and turn Available off or on.

TIP

Sending, Receiving, and Managing Emails on Apple Watch

Apple Watch would be a half-baked product if you could use it only for reading messages, not email. Fortunately, it delivers a decent mail experience right on your wrist. You can read and manage your mail, reply, and even compose a response, as shown in Figure 5-25.

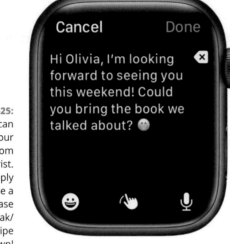

FIGURE 5-25: You can manage your inbox from your wrist. Tap to reply and choose a preset phrase or speak/ type/swipe your own!

Apple Watch's inbox is synced with your iPhone, so you can browse by date, sender name, titles, contents, and more. This feature is especially important for those who rely on email communication for work.

REMEMBER

Before you can read, reply, and manage your email on Apple Watch, you'll want to ensure that it's synced with the mail account(s) on your iPhone. Open the Apple Watch app on your iPhone, tap Mail, and ensure that your accounts are listed (see Figure 5-26). You can also tell your iPhone how you'd like to be notified of new emails on your Apple Watch by tapping Custom.

Reading and acting on an email

To read and act on an email message, follow these steps:

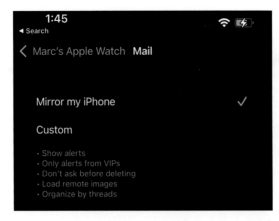
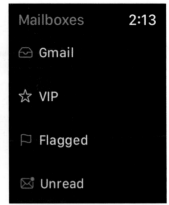

FIGURE 5-26:
The Mirror My
iPhone option
on the Apple
Watch app.
Make sure
it's enabled
in order to
receive email
on your wrist.

1. **If you're on the clock, press the Digital Crown button to go to the Home screen.**

2. **Tap the Mail app.**

   Or raise your wrist and say "Hey Siri, Mail." Either action launches the Mail app and takes you right to your inbox.

3. **Use the Digital Crown button or your fingertip to scroll up and down through your emails — as shown in Figure 5-27 — and then tap the subject line to open one.**

   The email you selected fills your Apple Watch screen, hiding the others.

4. **Scroll down to the bottom of the message and tap Reply, as shown in Figure 5-28.**

   You can also turn the Digital Crown button to scroll to the bottom. Press and hold Reply to see more options: Reply, Reply All (if others were included in this email), Mark as Unread, Archive (save for later), and Flag.

Replying to an email

If you want to reply to an email, tap Reply or Reply to All (if shown), and then you've got a few choices as to how to write your message, as shown in Figure 5-29:

>> **Typing:** This feature, available only for Apple Watch Series 7, Apple Watch Series 8, and Apple Watch Ultra, lets you pull up the optional QWERTY keyboard. Tap each letter with your fingertip, or swipe between letters on the virtual keyboard to use QuickPath.

>> **Scribble:** Reply by using the Scribble feature, which lets you use your finger to write a reply.

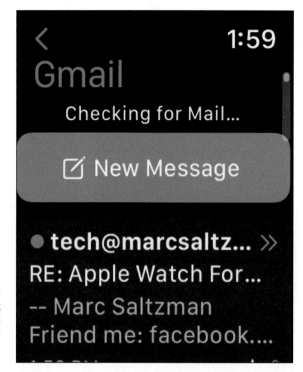

FIGURE 5-27:
Navigating
your inbox
by using the
Digital Crown
button.

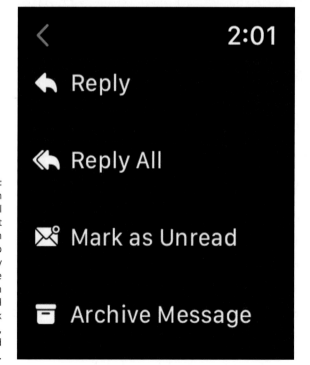

FIGURE 5-28:
Swipe down
in an email
message to get
to the bottom
and then tap
Reply, Reply
All (if anyone
else is in on
this email
thread), Mark
as Unread,
Archive, and
Flag.

>> **Preset Reply:** Select from one of the suggested replies offered by Apple Watch, such as "Let me get back to you" or "Got it, thanks." You can edit these phrases on your iPhone by tapping Apple Watch ➪ My Watch ➪ Mail.

>> **Microphone:** To dictate your email reply, tap this option (lower right of screen), which transcribes your voice as text. When you finish talking, tap Done. If the words onscreen look good, tap Send; if not, tap Cancel.

>> **Emojis:** Select the smiley face (in lower left of screen) to see many example emojis that you can send.

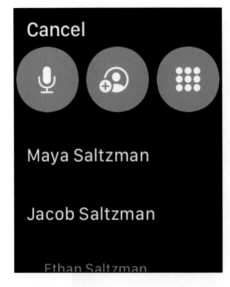 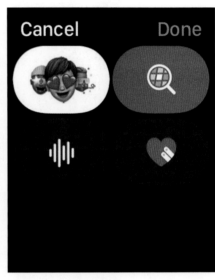

FIGURE 5-29: Tap Reply at the end of an email message, and you've got some options, including microphone (voice to text), Scribble, emojis, and default replies.

Composing an email

Although this feature wasn't available when the Apple Watch debuted, now you can create an email from scratch on the watch! Here's how to compose an email on Apple Watch:

1. **Open the Mail app.**

   You'll likely see emails in your inbox.

2. **Tap the words New Message.**

   You'll feel a slight vibration.

3. **Tap to add a Contact, Subject line, and Message (see Figure 5-30):**

   - For Contacts, choose the recipient from your Contacts list, or say a name, as shown on the left side of Figure 5-31.

- Tap Subject, and you have the option to dictate, Scribble, use an emoji, or choose among a list of preset Subject lines, as shown on the right side of Figure 5-31. If you have Apple Watch Series 7, Apple Watch Series 8, or Apple Watch Ultra, you can also type out a name or email address using the virtual keyboard.

4. **Tap Create Message.**

   You see options similar to those in Step 3. Plus, there are suggestions, which are recommended phrases, such as "Could you give me a call?"

5. **When you like what you see, tap Send.**

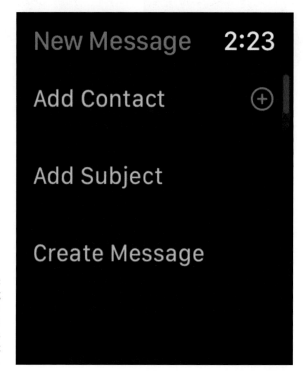

That's it!

Oh, wait, you say. What about Siri? Can't you raise your wrist and ask Siri to compose an email right from Apple Watch? Regrettably, no. If you try that, you'll hear Siri say that she can help you compose an email on your iPhone, if you like, but not on Apple Watch — or perhaps not yet.

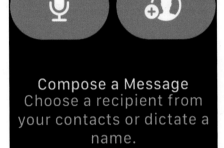

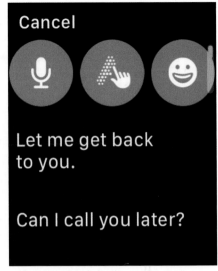

Cancel

Compose a Message
Choose a recipient from
your contacts or dictate a
name.

Cancel

Let me get back
to you.

Can I call you later?

Chapter **6**

In the Know: Staying Informed with Apple Watch

When people ask me "Why do I need a smartwatch?", I usually answer with something like "Convenience." A wearable device on your wrist is ideal for getting quick bits of information when and where you need information. Rather than pull out your smartphone or your tablet, you can simply glance down to get what you need while on the go.

In fact, when you start using Apple Watch, it's hard to go back to before you had it!

This chapter focuses mostly on how Apple Watch shows information that's relevant to you. You can access weather information for today, for the next week, or even for another city (or multiple cities). Do you like playing the market or want to see how your own public company is doing? Select which company's stock price and performance are important in your world — with all information updated in real-time.

The chapter also describes how to add this information to your watch face as one of the complications options and covers how to access notifications on Apple Watch for installed apps. I round everything off with directions for using Apple Watch for calendar alerts and accessing maps on your wrist.

Accessing Real-Time Weather, UV Index, and Wind Speed

Apple Watch comes with many built-in apps (see Chapter 1), and Weather is one of them. You can also add the Weather app to many watch faces. See Chapter 4 to find out how.

Sure, you can choose not to see this information on your smartwatch if you're not interested; simply deselect it in the Apple Watch app on your iPhone, as covered in Chapter 11. But this information should be timely and interesting for many people, which is why Apple provides it without requiring you to download specific apps.

Apple also added ultraviolet (UV) index info, wind speed, and (in some cities) the air quality index (AQI).

The Weather app displays your current location's temperature and a graphical representation of expected precipitation and conditions over much of the day (broken down by the hour) as well as a week-ahead view — for multiple locations, if you desire.

To use the Weather app on your Apple Watch, follow these steps:

1. **Press the Digital Crown button to go to the Home screen.**

2. **Tap the Weather app.**

   When the app launches, you should see the temperature and weather for your current location, as shown in Figure 6-1, which you can also add as a complication to a watch face. You can also see weather info for other locations you've selected in the iPhone version of the app.

FIGURE 6-1:
Get real-time
weather
conditions as a
"complication"
for a watch
face (per
Chapter 4).

3. **Tap your city.**

   You should see the temperature in the center of the app, surrounded by a 12-hour look at the day and an associated icon: a sun, a cloud, a raincloud, snowflakes, and so on. If you live in the United States, the temperature is displayed in Fahrenheit (a choice you make when setting up the watch); it's displayed in Celsius in Canada and other countries that use the metric system, or you can make that choice yourself. The name of the city appears in the top-left section of the screen, and the current time is in the top-right section.

4. **Swipe down or twist the Digital Crown button to scroll down the app.**

   You see a ten-day forecast, as shown in Figure 6-2.

   The further down you swipe or twist the Digital Crown button, the further ahead in the week you go. You can see each day's high and low temperatures and predicted precipitation — all courtesy of the Weather Channel. Scroll back up to the top of the Weather app by twisting the Digital Crown button, or use your fingertip to swipe up. At the top of the screen, you can see temperature, weather conditions, and precipitation; tap to change options like units of measurement, which cities to see, and more.

FIGURE 6-2:
The more
you scroll
down, the
more weather
information
you see.

5. **Keep scrolling to see the UV index and wind speed.**

6. **To see other cities you've selected to track, tap the top-left corner of the Apple Watch, which has a small arrow that enables you to select another city.**

   Tap the name of the city to see precipitation, too, and select what you prefer to see at a glance, like air quality, UV index, wind speed, and more, as shown in Figure 6-3. You can Customize what you see by using the Apple Watch app on your iPhone.

   You can change the cities you follow in the Apple Watch app on your iPhone; see Chapter 11 for more on doing this. You can select the default city or always see your current location (using location data in Apple Watch and/or the iPhone). You can also select the weather conditions you'd like to see on Apple Watch. Figure 6-4 shows examples of weather conditions you can access.

FIGURE 6-3:
Tap to access weather information in other cities, as well as wind speed, UV index, precipitation, and more.

REMEMBER

Don't forget that you can use Siri to call up weather information. Simply raise the watch to your mouth, say "Hey, Siri" (or press and hold the Digital Crown button), and ask a question, such as "What's it like outside?" or "Do I need a raincoat?" Or give a command, such as "Tell me the weather." Actually, you can just say "Weather." If you don't specify a location, as in these examples, Siri assumes that you want your local information. See Chapter 7 for more on using Siri to perform tasks with your Apple Watch.

Starting with watchOS 8, Apple Watch users can now receive "severe" weather advisories. When a significant weather event is predicted, a notification may appear at the top of the Weather app (see Figure 6-5). If you want to find out more about the government alert, tap Learn More.

Also new to the Weather app is "Next Hour" precipitation alerts.

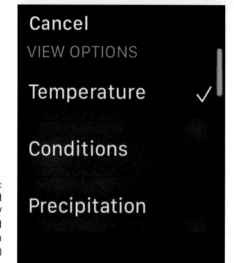

FIGURE 6-4:
watchOS 5 and
later have UV
index, wind
speed, and (in
some cities)
air-quality info.

FIGURE 6-5:
You'll see a
bolded
warning about
important
weather
events (and
emergencies)
in your
selected city,
which you can
tap to learn
more.

Following Stock Information and Much More

To help you keep an eye on the stock market, Apple Watch offers you a glance at any public company's stock price and performance when (and where) you'd like it. The Stocks app is similar to the Stocks app on the iPhone but tailored to the smaller Apple Watch screen. The process is similar to looking at the weather and just as customizable, including adding the app to a watch face (see Chapter 4).

To use the Stocks app on your Apple Watch, follow these steps:

1. **Press the Digital Crown button to go to the Home screen.**

2. **Tap the Stocks app.**

    If you didn't tweak the stocks you'd like to see in the Stocks area of the Apple Watch app on the iPhone, see Chapter 11 for more on doing that. Until you customize the stocks you want to keep an eye on, you see the current value of major stock indices, such as the Dow Jones Industrial Average, as well as such companies as Apple and Google (see Figure 6-6).

FIGURE 6-6:
See all kinds of
stock price and
performance
information
on any
publicly-traded
company.

Each company you follow on the stock market is listed by its traded name
(such as AAPL for Apple), with the current stock price below the name (such as
142.65), and whether the stock is up (in green) or down (in red) and by how
much, such as green +2.20%.

3. **Get more information about each company or exchange by tapping its
name (such as DOW J).**

The left side of Figure 6-7 shows an example. On the right, you can select the kind
of stock info to see, such as point and percentage changes and market cap.

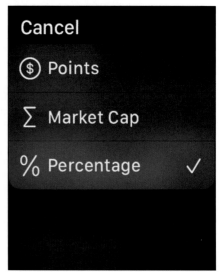

FIGURE 6-7:
Despite its
small screen,
Apple Watch
can give
you a ton of
information
about the
stock market,
such as this
snapshot of
the Apple
stock over the
past day.

You can see listings for several companies, all organized within the Apple Watch app on the iPhone. See Chapter 11 for more on third-party apps.

The iPhone Stocks app gives you an even deeper dive, such as the indices or company's performance over the past day, week, month, or six months (see Figure 6-8).

4. **Swipe to the right at any time to go back to the main Stocks app screen, which lists multiple indices and companies.**

You can also ask Siri to tell you the stock price of a given company or the performance of a stock index. You can access Siri by saying "Hey Siri," followed by your question or command, or press the Digital Crown button in any app you're in to ask Siri about a particular company.

FIGURE 6-8: View a snapshot of company stock or index performance over the past few hours (left). Or open the Stocks app on the iPhone for an in-depth view of the stock index or company, courtesy of Yahoo! (right).

Using Dock on Apple Watch

Covered briefly in Chapter 3, Dock lets you open and cycle through your most recently used apps (or your favorite apps) when you're on the go. In other words, it's a super-convenient shortcut to what matters most to you.

Launching Dock and more

Here's how to launch Dock, navigate among apps, launch one, and change what you see:

1. **Press the side button to activate Dock.**

2. **Swipe up or down with your fingertip, or turn the Digital Crown button.**

    This step cycles through the apps you opened most recently or your favorite apps. (The next section shows you how to customize this display.) Figure 6-9 shows what it looks like to cycle through apps.

FIGURE 6-9: Dock displaying recently used apps or your pinned favorites (up to ten).

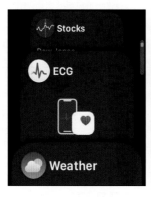  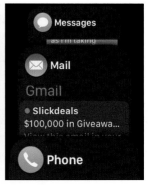

3. **Tap the name of an app to open it full-screen.**

    If you scroll all the way to the bottom of the screen, you can tap All Apps to go to the Home screen.

4. **To close an open app, swipe to the right and press the big red X.**

5. **To close Dock, press the side button again.**

Customizing Dock

To choose which apps appear in Dock (up to ten of your favorites), follow these steps:

1. **Grab your iPhone, and open the Apple Watch app.**

2. **Tap My Watch ⇨ Dock.**

   Here, you can choose your favorite apps.

3. **Tap Edit to add or remove apps (see Figure 6-10).**

   To remove apps, tap the red minus sign (–) and then tap Remove. To add apps, tap the green plus sign (+).

4. **To rearrange apps, touch and hold next to an app and then drag up or down.**

5. **Tap Done to save your changes.**

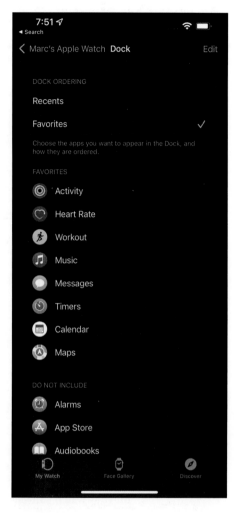

FIGURE 6-10:
In the iPhone Apple Watch app, you can select the apps to see when you activate Dock on your watch.

Mastering Notifications on Apple Watch

Just as many preinstalled Apple Watch apps support notifications (which are on by default), many third-party apps have them, depending on whether the developers provided them. This section shows you how to navigate notifications.

Seeing how notifications work

Notifications pull information from supported apps. If someone liked your photo on Instagram, if your favorite game wants you to know that someone is attacking your virtual kingdom, or if the New York Yankees won the double-header, Apple

Watch can notify you. Likewise, a news app such as USA TODAY might let you know when the president is about to give a speech or the weather is taking a turn for the worse in your area. Figure 6-11 shows an example notification.

FIGURE 6-11: USA TODAY and many other third-party news apps push headlines and other notifications to your wrist so you can see timely info that's meaningful to you.

A quick caveat: Notifications can go to either your Apple Watch or your iPhone, but not both. If your iPhone is unlocked, you'll get notifications on your phone rather than your watch. But if your phone is locked, asleep, or off, you get notifications on your watch (unless Apple Watch is locked with your passcode), as you can see in Figure 6-12.

If you have an Apple Watch without cellular connectivity, notifications are pushed to your watch from a nearby iPhone or over Wi-Fi. Otherwise, if you own a GPS + Cellular model and pay for the service, you can get notifications sent to your wrist directly from apps over the Internet.

Chapter 11 explains how to enable or disable app notifications.

REMEMBER

Accessing the notification settings

Quite simply, you can use the Apple Watch app on your iPhone to access notifications and choose which apps give you notifications. You can also select what information for which you'd like notifications. The right image in Figure 6-13 shows how to customize the Calendar app, which you do by tapping Custom. Notification Center also includes apps from third parties.

Thereafter, you should receive a gentle tap on your wrist when the apps you chose give you notifications.

TIP

If you're not feeling the notifications on your wrist, you can dial up extra vibration by tapping Settings ⇨ Sound and Haptics ⇨ Prominent Haptic on your Apple Watch.

Viewing notifications

When you receive a notification, such as a calendar appointment, a red dot appears on your watch face. As you do on your iPhone or iPad, swipe down from the top of the Apple Watch Home screen to access Notification Center. Scroll up or down with your finger or twist the Digital Crown button to see your next calendar appointment, find out how your stocks are doing, and perhaps get a look at traffic on the way to the office.

FIGURE 6-13:
In the Apple
Watch app on
the iPhone,
specify
whether you
want to get
notifications
on your watch
(left). You can
choose which
apps you'll
be notified
about (middle).
You can also
customize
notification
information
(right).

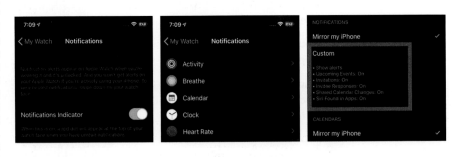

FIGURE 6-13: In the Apple Watch app on the iPhone, specify whether you want to get notifications on your watch (left). You can choose which apps you'll be notified about (middle). You can also customize notification information (right).

To view notifications on your Apple Watch, follow these steps:

1. **From the watch-face screen, press or swipe down from the top of the screen to open Notification Center.**

   Start swiping down from the very top of the watch case — on the rim — to pull up the information you want. Swiping from the middle of the screen doesn't work. Also, if you press too close to the center of the watch face, you pull up all your watch faces rather than notifications. Finding the sweet spot may take a bit of trial and error.

   Depending on the app, you may first see a summary of the notification and may need to keep your arm raised to see more details.

2. **Swipe up and down for additional information, or twist the Digital Crown button.**

   To go back to your Home screen, press the Digital Crown button.

3. **View and then dismiss notifications by swiping up.**

   When you dismiss notifications by swiping up, they're also dismissed from your iPhone. You can also delete a notification by swiping to the left and tapping the large *X*.

Changing how you get notifications on your Apple Watch

Apple Watch isn't a one-size-fits-all scenario. Different wearers will want to be notified about different apps and in different ways. Here's how to make changes:

1. **Press the top of the watch face or swipe down from the top of the watch face to open Notification Center.**

2. **When Notification Center appears, swipe down.**

3. **Swipe left on a notification, tap the three little dots, and then select an option (see Figure 6-14):**

   - *Deliver Quietly:* If you don't want to hear sounds or feel haptic alerts for that app, tap Deliver Quietly. Thereafter, notifications will go directly to Notification Center on both your Apple Watch and iPhone without producing a sound or haptic alert.

   - *Turn Off on Apple Watch:* If you don't want to get notifications for an app, tap this option.

4. **If you change your mind about an app, swipe left on a notification from the app, tap the three dots, and then choose Deliver Prominently.**

FIGURE 6-14:
Choose how to receive app notifications.

TIP

When you raise your wrist to view a notification, you see a quick summary and then full details a few seconds later. Want more privacy? To prevent the details from appearing in the notification, open the Apple Watch app on your iPhone, and tap My Watch ⇨ Notifications ⇨ Notification Privacy. Now when you receive a notification, you need to tap it to see the full details.

Accessing Your Calendar on Apple Watch

Because your Apple Watch is connected to the Internet via cellular or Wi-Fi — or at least wirelessly tethered to your nearby iPhone through Bluetooth technology — you can access handy calendar information on your wrist. In fact, the Calendar app syncs not only with your iPhone, but also with iCloud (if you use Apple's popular cloud service to store and access information). This section covers how to add an event to the Calendar, navigate the Calendar app, respond to a calendar or appointment request, and reply to an appointment request through notifications.

Navigating the Calendar app

Featuring day, week, and month views — including support for reminders, invitations, and notifications — the Calendar app on Apple Watch shows you a list of upcoming events.

To use the Calendar app on your Apple Watch, follow these steps:

1. **Press the Digital Crown button to go to the Home screen.**

2. **Tap the Calendar app.**

   The Calendar app launches. The first time it opens, you need to allow the app to know your location, as shown in Figure 6-15. Why? If an event includes a location, you automatically get a "leave now" alert on your Apple Watch, based on estimated travel time and traffic conditions! Your options are Allow Once, Allow While Using App, and Never.

   By default, you should see the current Today view, with your upcoming events listed in chronological order (see Figure 6-16).

3. **Use your fingertip to scroll down to see future dates, or twist the Digital Crown button toward you.**

   This experience is similar to using the Calendar app on iPhone. The current time is listed in the top-right section.

4. **In Today view, tap the top-left corner to access Month view; to return to Today view, tap a specific date.**

   Figure 6-17 shows Month view. In this view, you can swipe left or right to move forward or backward through time, or twist the Digital Crown button if you prefer.

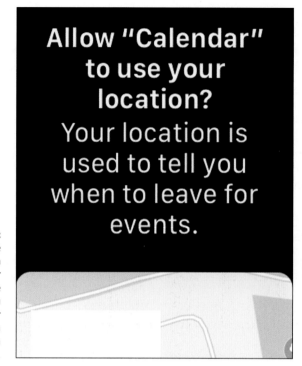

FIGURE 6-15:
The first time you open the Calendar app on Apple Watch, you share your location information with it.

View a monthly calendar.

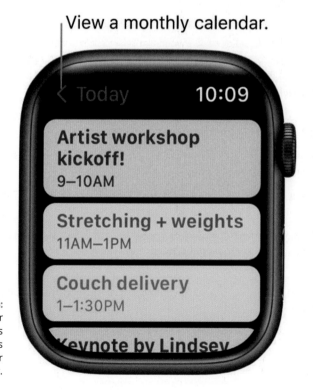

FIGURE 6-16:
The Calendar app shows the events scheduled for your day.

TIP

Just as you can set Reminders on your iPhone, iPad, or iPod touch, you can activate Siri on your Apple Watch and say something like "At 6 p.m., remind me to call Mom." The watch doesn't make a calendar entry for this event, but you're reminded with a sound, a vibration on your wrist, and a text. Apple Watch has no keyboard, so you must dictate the reminder.

5. **Create a calendar appointment on your Apple Watch.**

    There are two ways to do this:

    For one, you'll need to use Siri. Simply ask your personal assistant to make a calendar entry on a given date and time. (See Chapter 7 for more ways to use Siri to accomplish tasks with Apple Watch.)

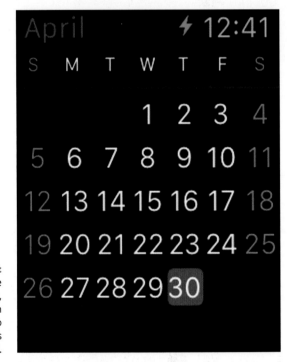

FIGURE 6-17: From the Month view, just tap on any day to see what's scheduled.

Starting with watchOS 9, the second way is to tap the large "+Add Event" option on your Calendar app. See Figure 6-18.

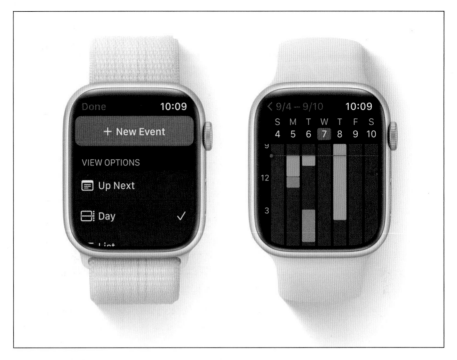

FIGURE 6-18:
It's now easier
to create new
events directly
from your
Apple Watch
and to navigate
to specific days
or weeks. Sure,
you can still
use your voice
via Siri, but
now there's an
Add Event tab
to select,
as well.

Responding to a calendar/appointment request

Suppose that you received an invite or are reviewing one. Apple Watch makes it easy to reply without your iPhone. Also, you can perform other tasks, such as sending an email quickly or sharing map info. Follow these steps:

1. **Tap a calendar invite, as shown in the top-left image of Figure 6-19.**

   You see the date and time of the appointment, as well as the calendar/email account it's tied to (in case you have different accounts).

2. **Press and hold the screen to see more info about the calendar event, such as any details or notes associated with it.**

3. **Scroll down and tap Accept, Decline, or Maybe (top-right and bottom-left images in Figure 6-19).**

4. **Press and hold the name of the organizer.**

   Other options (bottom-right image of Figure 6-19) appear, including sending the invitee an email or getting directions to where you're going. You can also get the estimated driving time from your current location via Apple Maps. (See "Navigating the Maps App" later in this chapter.) You can also send the organizer a message, if you like, right from the Calendar app on Apple Watch.

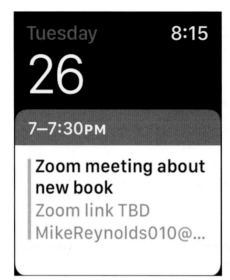
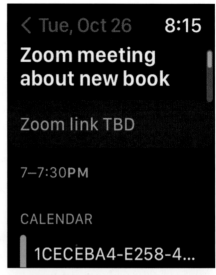
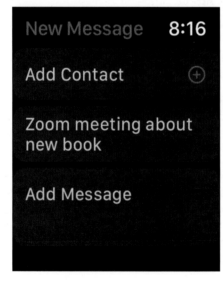

FIGURE 6-19: Responding to a calendar request is as easy as selecting it, accepting it, and sending the sender a confirmation email.

Don't forget that Apple Watch pulls calendar events from your iPhone. Also, you don't always have to check your calendar for upcoming appointments, because you should receive a notification about it (and feel a slight pulse). Some users like to be reminded an hour before an event, for example, whereas others want a five-minute reminder. This setting is in your iPhone's Calendar app.

Accepting a calendar request through notifications

You can set your Apple Watch to receive calendar invitation notifications — via the Apple Watch app on the iPhone — which you can accept or decline immediately. This section shows you how to accept or decline a calendar invitation or to reply to the organizer from your Apple Watch.

As a setup, assume that someone sent you a calendar invite via email or message. You should receive a notification with the proposed meeting date, time, and information, such as "Natalie's B-Day Party, December 1, 2022, 6 p.m." Follow these steps to respond:

1. **Swipe down on the notification or twist the Digital Crown button to open Notification Center.**

2. **Scroll down and tap Accept, Maybe, or Decline.**

   You can't suggest an alternative date or time, as you can with some email programs, but you can give the organizer a response by dictating a reply (or with Apple Watch Series 7 or 8, or Apple Watch Ultra pulling up a QWERTY keyboard, if desired).

3. **Add the event to your calendar.**

   You don't need your iPhone to receive calendar alerts on your Apple Watch. Therefore, if you go on a run or accidentally leave your iPhone at the office, you can still see existing calendar entries.

   Don't forget that you can raise your wrist or say "Hey Siri" — or press and hold the Digital Crown button — and then say something like "Add calendar entry, dentist appointment, for 9 a.m. tomorrow." This spoken text is added to your calendar and synced with your iPhone too. (See Chapter 7 for more ways to use Siri with your Apple Watch.)

Setting Reminders on Apple Watch

Since watchOS 6, Apple Watch has made it easy to set reminders and get notifications about them at the right time (or in the right place). If you use the Reminders app on the iPhone and iPad (running iOS 13 and later) and/or the app in macOS Catalina or later, the Reminders app on Apple Watch will be familiar.

To see and set reminders on Apple Watch, follow these steps:

1. **To see your reminders, tap the white Reminders icon on the Apple Watch Home screen.**

   Or raise your wrist, say "Hey Siri," and then say something like "Open reminders" or "Do I have any upcoming reminders?"

2. **Tap one of the sections to review your reminders: Today, All, Scheduled, and Flagged (see Figure 6-20).**

   You'll be notified of a reminder at the time you specified in Settings via a slight vibration on your wrist, an alarm sound, or both (you can select in Settings).

   All your reminders are conveniently synchronized among your Apple devices unless you don't want them to be (which you can tweak in the Watch app on your iPhone).

3. **To set a reminder in the Reminders app, scroll down or turn the Digital Crown button, select Add Reminder, and then use the Speech (voice to text) or Scribble (handwriting to text) feature to create the reminder.** Or, on Apple Watch Series 7, Apple Watch Series 8, or Apple Watch Ultra, you can use the QWERTY keyboard, too. Chapter 5 discusses all the ways to input text on Apple Watch.

   If you see the Reminder notification when it arrives, you can swipe the screen to dismiss it or turn the Digital Crown to scroll to the bottom of the reminder, and then tap Snooze, Completed, or Dismiss.

   If you discover the notification later: Tap it in your list of notifications, and then scroll and respond.

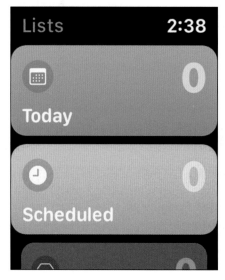
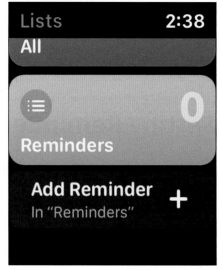

FIGURE 6-20: With the Reminders app, you can review your reminders (by Day, All, Scheduled, and Flagged) or create a new one.

Don't forget to leverage Siri as the fastest way to create reminders on Apple Watch. So long as you're in a place where you can talk aloud, raise your wrist; say "Hey Siri:" and ask for a reminder, such as "Tomorrow at 9 a.m., remind me to email my book editor" or "When I get to the office, remind me to create new passwords." Chapter 7 covers Siri in depth.

Accessing Apple Watch's Integrated Calculator

Apple Watch puts a handy calculator on your wrist. Although this calculator isn't super-high-tech — heck, I had a Casio calculator wristwatch in the '80s! — it sure is convenient to have an integrated calculator app on a watch.

To use the built-in calculator, tap the Calculator app icon on your Apple Watch Home screen. Then you can perform a math operation, including addition, subtraction, division, and multiplication (left side of Figure 6-21). Tap AC ("All Clear") to clear the number.

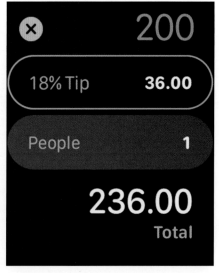

FIGURE 6-21: The Calculator app for Apple Watch.

Alternatively, you can ask Siri to open the Calculator app (or ask Siri your math question).

If you want your Apple Watch to generate a tip if, say, you're at a restaurant, type the amount of the bill (such as $120), and the watch generates the tip. (The default setting is 15%, but you can turn the Digital Crown button to set a higher or lower number.) You can also select how many people are splitting the bill so you'll know how much each person owes!

Creating and Listening to Voice Memos

Sometimes it's more convenient to use your voice than type information. You can use the dedicated Voice Memos app on the Apple Watch Home screen, or you can add a Voice Memo complication (see Chapter 4) to your favorite watch face so you can tap and leave yourself a voice recording via the watch microphone. If wireless earbuds or headphones are connected to Apple Watch, you hear the voice recordings there, not through the watch itself.

Tap the Voice Memos app to open it or raise your wrist and instruct Siri to open the app ("Open Voice Memos"). Then you can perform the following tasks:

>> **Record a voice memo.** Tap the large red Record button, shown in Figure 6-22, and begin talking to your wrist.

>> **Stop recording a voice memo.** To stop the recording, tap the red Stop button.

>> **Play back a voice memo.** Tap its name below the Red record button. If you didn't name the recording, it's listed as something like Recording 1 or Recording 2. Press Pause to pause playback or skip forward or back 15 seconds.

>> **Edit the name of the recording.** Tap the name (such as Recording 3), and you have the option to use your voice or the Scribble feature to create a custom name.

>> **Erase the voice memo.** Tap the three little dots in the bottom-right section of the screen to erase the voice recording. You're asked to confirm your decision to erase.

Recordings are also added to the Voice Memos app on your iPhone, iPad, and Mac.

If you press and hold the screen inside the Voice Memos app on Apple Watch, you can play back recordings stored on the Apple Watch itself (in a section called Watch Recordings) or tap All Recordings to hear recordings from other Apple devices.

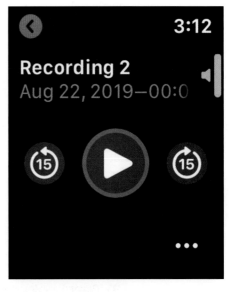

FIGURE 6-22:
You can leave
yourself handy
voice notes
through your
Apple Watch
and play them
back whenever
you like.

Navigating the Maps App

Apple Maps — or simply Maps — is a standard Apple Watch app that allows you to get directions for the best route from your current location to a destination of your choice. Because all Apple Watch models have integrated GPS (except for the first model from 2015), the app talks with satellites to determine your exact location. Think about that for a moment: Your wristwatch is talking to satellites in space. Crazy!

Pairing GPS with mapping software means that you can easily find your destination or share your location. What's more, a Compass app is available on Apple Watch Series 5 and later, and Apple Watch SE has a Compass feature, which can also help you navigate accurately. More on this shortly.

When you're en route somewhere, you should see — and feel — turn-by-turn navigation instructions that guide you along the way, and you can always search for nearby businesses, such as a restaurant or a gas station, simply by asking Siri.

To use the Maps app on your Apple Watch, follow these steps:

1. **Press the Digital Crown button to go to the Home screen.**

2. **Tap Maps.**

   The Maps app launches. Depending on where you left the app last, you see one of two options on your Apple Watch screen (see Figure 6-23):

   - *Search:* Tap the empty search field to type in a location using the onscreen keyboard (Apple Watch Series 7, Apple Watch Series 8, or Apple Watch Ultra) or tap the little microphone to verbally mention a place, such as a restaurant or actual address.

   - *Location:* Tap this, and you'll see an overhead map of your current location: Swipe to move the map around or twist the Digital Crown button if you want to see nearby streets or businesses.

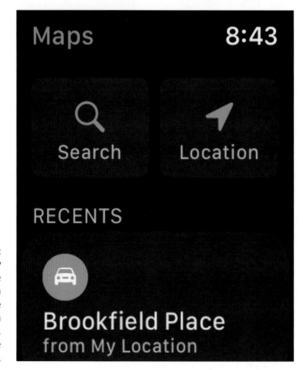

FIGURE 6-23: You may see these options when launching the Maps app on Apple Watch. Choose wisely, Jedi.

3. **In Maps view, tap the blue icon in the bottom-left section of the screen (see Figure 6-24) to return to your current location.**

   This action recenters the map on your specific location. You can also zoom in and out by twisting the Digital Crown button.

FIGURE 6-24:
The Maps app offers features that help you with directions or find local businesses.

4. **Find a location in Maps view by tapping the three dots in the lower right corner of the screen.**

   You're given two options:

   - *Search Here:* Select this option (see Figure 6-25) to search for something nearby, such as a coffee shop or dry cleaners. You're prompted with other options. You can search by Dictation (use your voice to say what you're looking for and want directions to), use Scribble (write a word with your fingertip), or select Contacts to find someone in your address book. Or you can pull up information divided into the categories Food, Shopping, Fun, and Travel. Whatever you select, the app pulls up information about relevant businesses near your location.

   - *Transit Map:* Tapping this option allows you to view a new map overlaid with bus, train, and/or subway information. You can tap the colored line to pull up additional information, as shown in Figure 6-26.

TIP

The next time you search in Maps, you should also see your last searched addresses, as mentioned in step 2.

FIGURE 6-25:
The Map
screen's Search
Here option
(left) allows
you to search
with Dictation,
Scribble
(middle), or
Contacts (right)
when you're
lost or to pull
up information
on nearby
establishments
(below).

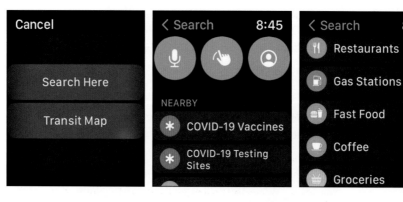

FIGURE 6-26:
Apple Maps
has transit
maps for
bus, train,
and subway
information
(left). Tapping
a transit
option pulls
up additional
details and
directions.

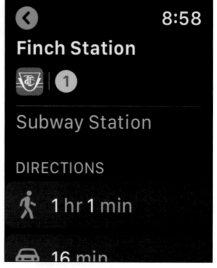

5. **Tap a location you want to travel to, such as a sushi restaurant.**

If the location is a business, you can tap it on the map to bring up information such as its address and phone number (which you can call), hours of operation, and star rating on Yelp (average user rating out of five stars). You should also see an estimate of how long it will take to get there by foot or by car (see Figure 6-27).

6. **Tap the mode of transportation you're using.**

You should see the destination as a pushpin on the map.

If you own an Apple Watch Series 5 or later, look at the little compass dial in the bottom-left corner of the Maps screen to see where north is.

TIP

Addresses in Messages, Email, Calendar, and other apps are highlighted in blue and underlined, which means that they're tappable. Tapping an address, such as 123 Yonge St., launches that address in the Maps app for you. Neat, huh?

7. **Tap Start to map your route.**

   You're off!

8. **Follow Apple Watch's instructions as you make your way to your destination.**

   If you need to turn right, you feel a steady series of a dozen taps on your wrist at the intersection you're approaching. To turn left, you feel three pairs of two taps. You can also look at your screen for visual cues (while you're walking, not driving, of course!).

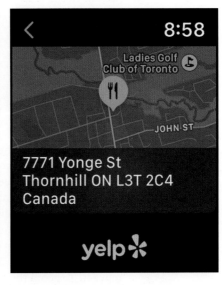

FIGURE 6-27: Get information about a local business provided by Yelp by tapping its name.

Drop, move, and remove map pins

Just like you can with Apple Maps on iPhone and iPad, you can drop a virtual red pushpin on a map to mark the destination:

1. **Touch and hold the map where you want the pin to go, wait for the pin to drop, then let go.**

   To move a pin, drag it around on the map.

2. **To remove a pin, tap the pin to see address information, turn the Digital Crown to scroll, then tap Remove Marker.**

3. **To find the approximate address of any spot on the map, drop a pin on the location, then tap the pin to see address info! See Figure 6-28.**

4. **Tap how you will be traveling there, such as walking, driving, transit, or cycling.**

FIGURE 6-28:
To drop a pin, press and hold somewhere on the map. Tap the red dot again to launch the directions there.

As you can with all other built-in apps, use Siri to access Maps information whenever you like. Raise your wrist and say "Hey Siri," followed by something like "Show me 5 Main Street in Beverly, Kansas" to see it on a map or "Give me directions to the Golden Gate Bridge." You can also press the Digital Crown button to activate Siri. (See Chapter 7 for more on using Siri with your Apple Watch.)

TIP

You can pan and zoom on the Apple Maps app on Apple Watch — just like you can on an iPhone or iPad. To pan the map, simply drag one finger across the map to move it. To zoom the map in or out, turn the Digital Crown. You can also double-tap the map to zoom in on the spot you tap. To go back to your current location, tap the Location button at the bottom left.

Getting Directions on Apple Maps

Okay, so you decided to visit a friend out of town who moved into a new place. You raise your wrist and instruct Siri to take you to the address.

This opens up Apple Maps and shows you the destination. Now what?

Just like Apple Maps on iPhone, you'll select how you're going to get there. For example, choose Driving. Now it's time to hit the road.

1. **Before you drive, tap a route to begin your trip, and you'll see an overview of it, which includes turns, distance between turns, and street names.**

2. **Look at the top-left corner of your watch to see your estimated arrival time (ETA).**

   The Maps app also shows the name of the street where you next turn as well as the distance before you make that turn.

3. **When safe to do so (such as when not driving):**

   - When viewing a list of turn-by-turn directions, tap the Map button to open a map that shows the turn's location. Turn the Digital Crown to zoom in and out on the map. Tap the List button to return to the turn-by-turn list.

   - Turn the Digital Crown to see upcoming turns, then tap the top of the display to return to the next turn you'll take.

Note: Location services must be turned on to use turn-by-turn directions. On Apple Watch, go to Settings ⇨ Privacy ⇨ Location Services and make sure that location services are enabled (green).

On your journey, listen for directions! Apple Watch uses sounds and taps to let you know when to turn. A low tone followed by a high tone ("tock tick, tock tick") means turn right at the intersection you're approaching; a high tone followed by a low tone ("tick tock, tick tock") means turn left. You'll feel a vibration when you're on the last leg, and again when you arrive.

To end directions before you get there, tap the End button in the bottom right of the watch screen.

Using the Compass

If you own Apple Watch Series 5 or later or Apple Watch SE, a built-in compass that points north inside the Maps app helps you get where you're going, thanks to an integrated sensor called a *magnetometer* that detects levels of magnetism. In other words, the compass works even when you don't have Wi-Fi or a cellular connection! As shown in Figure 6-29, the little blue compass is in the bottom-left corner of the screen.

Figure 6-30 shows how you can add a Compass complication to most watch faces on Apple Watch in case you like seeing this information all the time.

FIGURE 6-29: The updated Maps app shows the direction in which you're facing.

FIGURE 6-30: A Compass complication (bottom middle) for several watch faces shows directional information at a glance.

You can go into the stand-alone Compass app for a deeper dive into magnetometer info, displaying your heading along with elevation, latitude, longitude, and incline (see Figure 6-31).

Introduced in Apple Watch Series 6 (2020), an always-on altimeter in the Compass app can give you real-time elevation data as well. You can add the Elevation complication to select watch faces that show your current elevation (see Figure 6-32).

To add the Elevation complication to a watch face, follow these steps:

1. **Press and hold the watch-face screen; then tap Edit at the bottom.**

2. **Swipe all the way left to access complications.**

    See Chapter 4 to find out how to work with complications.

3. **Tap a complication area to select it, turn the Digital Crown button to Compass, and then choose Elevation.**

4. **Press the Digital Crown button to save your changes.**

5. **Tap the watch face to switch to it.**

FIGURE 6-31:
Lost in the woods? You might be fine, thanks to the Compass app on an Apple Watch Series 5 or newer and Apple Watch SE.

Note: Compass won't display elevation or coordinates if Location Services is turned off. To turn Location Services on or off, tap Settings ⇨ Privacy ⇨ Location Services on your Apple Watch.

To use true north rather than magnetic north, tap Settings ⇨ Compass on your Apple Watch, and turn on Use True North.

Apple cautions you to be aware of the presence of magnets, which will likely affect the accuracy of the compass's sensor. In fact, some of the Apple Watch bands — including Leather Link, Leather Loop, Milanese Loop, and earlier Sport Loop bands — use magnets (or magnetic material) that might interfere with the compass. (The compass won't be affected by Sport Loop bands introduced in September 2019 or later or by any version of the Sport Band.)

FIGURE 6-32:
Because of
the always-on
altimeter in
Apple Watch
Series 6, you
can add real-
time elevation
data to many
watch faces.

Using the Find People App

The Find My app on your iPhone helps you find friends and family members, and perhaps share your location with them. The Find People app on Apple Watch works in a similar fashion.

So long as your peeps are willing to share this info with you via their Apple devices (iPhone, iPad, iPod touch, Apple Watch SE, or Apple Watch Series 3, or later), you can see and share their locations on a map, or set notifications to alert you when those who matter leave from or arrive at various locations. See Figure 6-33.

Using the Find People app is a breeze. The following sections explore what you can do with it.

Adding a friend to the Find People app

To add someone, follow these steps:

1. **Open the Find People app on your Apple Watch.**

2. **Tap the blue plus sign (+) next to Share My Location.**

FIGURE 6-33:
As the name suggests, the Find People app for Apple Watch helps you find people via their compatible Apple devices and share your location too.

3. **Tap the Dictation, Contacts, or Keypad button to choose a friend.**

4. **Choose an email address or phone number.**

5. **Select how long to share your location: one hour, until the end of the day, or indefinitely.**

Now your family member or friend will receive a notification that you've shared your location. If they want, they can choose to share their location too. After the person agrees to share their location, you can see where they are on a map or in a list in the Find My app on an iPhone, iPad, iPod touch, Mac, or Apple Watch.

To stop sharing your location with someone at any time, tap your friend's name on the Find People screen and then tap Stop Sharing.

Alternatively, to stop sharing your location with everyone, tap Settings ⇨ Privacy ⇨ Location Services on Apple Watch and then turn off Share My Location.

Finding out where your friends are

Want to know where your gang is? Easy-peasy. Follow these steps:

1. **Open the Find People app on Apple Watch to see a list of your friends who've agreed to share their locations with you.**

   To see more friends, turn the Digital Crown button.

2. **Tap a friend's name to see their location on a map (see Figure 6-34), or ask Siri "Hey Siri, where is Ashley?"**

3. **To return to your friends list, tap the little < symbol in the top-left corner.**

FIGURE 6-34: Apple Watch's Find People app makes it easy to find friends and family — so long as they're on a compatible Apple device and willing to share their location with you.

Notifying a friend of your departure or arrival

You can let people know when you're leaving or arriving somewhere by following these steps:

1. **Open the Find People app.**

2. **Tap a friend's name, scroll down, and tap Notify [name of friend].**

3. Activate Notify [name of friend] on the next screen and choose to notify your friend when you leave your location or arrive at their location.

Getting a notification about a friend's location

On the flip side, you can ask to be notified of someone's location by following these steps:

1. Open the Find People app.

2. Tap your friend's name, scroll down, and tap Notify Me.

3. Activate Notify Me and choose to be notified when your friend leaves their location or arrives at your location.

Locating Your Stuff with the Find Items App

As you might expect, you can use your Apple Watch to locate *AirTags*, which are small Bluetooth trackers released in 2021 (see Figure 6-35). Clip AirTags to items such as car keys, a dog collar, a purse, a backpack, and a TV remote.

FIGURE 6-35: AirTag is a small accessory that can be personalized with free engraving, and enables Apple Watch and iPhone users to locate valuables it's connected to.

How does this feature work? Apple's AirTags broadcast a Bluetooth signal to help you locate lost items. When the tracker is within range of your iPhone or Apple Watch (around 30 feet), you can locate the item quickly by using the Find My app (iPhone) or Find Items app (Apple Watch). The tracker also emits a sound when you tap to find the item. You can also ask Siri to find your missing item; the AirTag will play a sound if the item is nearby.

AirTags use ultra-wide-band technology to lead you to the items' locations. If you're on an iPhone with a U1 chip (iPhone 11 family and later, excluding iPhone SE devices), a directional arrow points you directly to an AirTag's location (and shows how far away it is in real-time). Your location data and history are never stored on the AirTag itself, Apple says.

Here's where AirTags get really interesting. If you left your backpack at a friend's house — therefore, out of Bluetooth range — the Find My network may help you track it down. Using roughly 1 billion Apple devices around the globe, it can detect Bluetooth signals from an AirTag and relay the location back to you. Also, you can place AirTag in lost mode in the Find My app and be notified when it has been located. If someone finds your stuff, they can tap it on their iPhone or any other device that's capable of near field communications (NFC), and they'll be taken to a website that shows how to reach you (which you set up ahead of time when you register an AirTag on your iPhone).

AirTags have a user-replaceable battery that lasts about a year.

"Okay, but how do I use my Apple Watch to find something?" you ask. The first step is setting up the AirTags on your iPhone. If you don't know how, Apple has an online tutorial; search for "how to set up AirTag on iPhone" on the web.

The following sections show you how to use your Apple Watch to find something.

Seeing the location of an item

Open the Find Items app on Apple Watch and then tap an item you want to find (see Figure 6-36).

Keep the following things in mind:

>> **If the item can be located,** it appears on a map, which shows the device's approximate location and distance from you, the time it last connected to Wi-Fi or cellular networks, and its charge level.

>> **If the item can't be located,** you see where and when it was last located. Tap Notifications and turn on Notify When Found to receive a notification when the device is located again. Make sense?

FIGURE 6-36:
Locate a
missing item
in your Apple
Watch's Find
Items app.

Marking an AirTag as lost

If you lose an AirTag, you can use the Find Items app on Apple Watch to mark it as lost. Follow these steps:

1. **Open the Find Items app.**

2. **Tap the item, such as Alan's Backpack.**

    You should see a description where your missing item was last seen (see Figure 6-37) and an optional map, in the hopes you can retrieve it.

3. **Activate Lost Mode.**

Did you get your missing item back? To turn off lost mode for an item, follow these steps:

1. **Open the Find Items app.**

2. **Tap the item.**

3. **Deactivate Lost Mode.**

Making an AirTag play a sound

If the item is nearby, you can make it play a sound to help you find it. Follow these steps:

1. **Open the Find Items app.**

2. **Tap the item you want to play a sound on.**

3. **Tap Play Sound.**

4. **To stop playing the sound before it ends automatically, tap Stop Sound.**

Getting directions to an item

It's possible to see directions to an item's current or last known location in the Maps app on your Apple Watch. Here's how:

1. **Open the Find Items app.**

2. **Tap the item you want to get directions to (such as Kellie's Purse).**

3. **Tap Directions to open Maps.**

4. **Select the route to get directions from your current location to the item's location.**

Receiving a notification when you've left an item behind

Maybe you're the forgetful type, and you want to know when you leave something behind. As long as an AirTag is connected to an item, this notification is easy enough to set up. Follow these steps:

1. **Open the Find My app.**

2. **Tap the item you want to set up a notification for.**

FIGURE 6-37:
Open up the AirTag app on iPhone to mark the item as lost, and you'll see on your iPhone where it was last located.

3. **In the Notifications section, activate Notify When Left Behind.**

4. **Follow the onscreen instructions.**

If you want to add a trusted location, you can choose a suggested location or tap New Location, select a location on the map, and then tap Done.

TIP

If you want to change distance units, open the Find Items app on your Apple Watch, turn the Digital Crown button to scroll to the bottom of the screen, tap Directions In, and choose miles or kilometers.

Finding misplaced devices via the Find Devices app

It's the worst feeling when you realize that you can't find your iPhone, iPad, iPod touch, AirPods, Apple Watch, or Mac. But if you own an Apple Watch, you may be able to locate these lost devices thanks to the aptly named Find Devices app, which was added to watchOS 8.

REMEMBER

To use this feature, you need to be signed in to the missing device with your Apple ID.

Seeing the location of a device

If your device is online, you should be able to see its location in the Find Devices app. But even if it's powered off or in low-power or airplane mode, you may still be able to find it. More on this in a moment.

To see a device's location, open the Find Devices app on your Apple Watch and then tap the name of the device (see Figure 6-38).

What happens next is tied to where you last left the item:

>> **If the device can be located (such as when it's online) and maybe in your home,** it appears on a map, which shows the device's approximate location and distance from you, the time it last connected to a Wi-Fi or cellular network, and its charge level.

>> **If the device can't be located,** you see the words **No location** below the device's name. Tap Notifications and turn on Notify When Found to receive a notification when the device is located.

Playing a sound on your device

Maybe your iPhone is stuck between the sofa cushions. You can have it emit a loud sound that will help you locate it. Follow these steps:

1. **Open the Find Devices app.**

2. **Tap the name of the device.**

3. **Tap Play Sound. It makes the AirTag ring, but you don't get a choice in sounds.**

If the device is online, a sound starts after a short delay and gradually increases in volume, playing for a total of two minutes. The device also vibrates (if it supports this feature). A Find My [device] alert appears on the device's screen. A confirmation email is also sent to your Apple ID email address.

If the device is offline, you see the words *Sound Pending*. The sound plays the next time the device connects to a Wi-Fi or cellular network.

Getting directions to a device

You can get directions to a device's current location in the Maps app on your Apple Watch. To get going, follow these steps:

1. **Open the Find Devices app on your Apple Watch.**

2. **Tap the name of the device.**

3. **Tap Directions to open Maps.**

4. **Tap the route to get directions from your current location to the device's location.**

Receiving a notification when you've left a device behind

You can opt to receive a notification if and when you've left a device behind. You can also set trusted locations where you can leave your device without receiving a notification. To use this feature, follow these steps:

1. **Open the Find My app on your iPhone.**

2. **Tap Devices.**

3. **Tap the device you want to set up a notification for.**

4. **In the Notifications section, activate Notify When Left Behind.**

5. **Follow the onscreen instructions.**

TIP

Alternatively, you can open the Find Devices app on your Apple Watch, tap a device, scroll up, and tap Notify When Left Behind.

Marking a device as lost

If your device is lost or stolen, you can turn on lost mode for your iPhone, iPad, iPod touch, or Apple Watch, or lock your Mac. Here's how:

1. **Open the Find Devices app.**

2. **Tap the name of a device.**

3. **Tap Lost Mode.**

When you mark a device as lost, the following things occur:

>> A confirmation email is sent to your Apple ID email address.

>> A message indicating that the device is lost and how to contact you appears on the device's lock screen.

>> Your device doesn't display alerts or make noise when you receive messages or notifications, or when alarms go off. Your device can still receive phone calls and FaceTime calls.

>> Apple Pay is disabled for your device, and any credit or debit cards set up for Apple Pay are removed from the device.

>> For an iPhone, iPad, iPod touch, or Apple Watch, you see your device's current location on the map as well as any changes in its location.

TIP

If your device was stolen, and you see where it is on a map (such as someone's home), never try to retrieve it on your own. Instead, contact the authorities to handle the situation. Why put yourself in danger's way?

3

It's All in the Wrist

Gain a helping hand from Siri, which can assist you in completing tasks with your Apple Watch, including tips and tricks to speed up your requests.

Get physical with help from the Activity and Workout apps on your Apple Watch, including setting and modifying goals and earning rewards for your achievements. A new Trends feature shows whether any metric is headed up or down over time, so you can keep it going or turn it around.

Learn all about the Mindfulness app, how to access the electrocardiogram (ECG or EKG), blood O2 monitor, fall detection, and SOS features (including Crash Detection), and use the Cycle Tracking app.

Access music, podcasts, audiobooks, and radio plays on your Apple Watch and take your favorite media with you anywhere.

Turn your Apple Watch into a virtual wallet by setting up Apple Pay. Then explore the Wallet app, where you can store movie and concert tickets, boarding passes, coupons, loyalty cards — and even use your Apple Watch as a car or house key.

Chapter **7**

Siri Supersized: Gaining the Most from Your Personal Assistant

Although many Apple Watch wearers will interact by using their fingers on their wrist-mounted gadget — tapping, pressing, or swiping the screen or accessing the two buttons along the side — you can get more done in less time if you simply talk to your watch.

Already an iconic feature on other Apple products (iPhone, iPad, Mac, HomePod, and Apple CarPlay–enabled vehicles) Siri (pronounced "sear-eee") is your own voice-activated personal assistant. Using your words rather than your fingers to ask for information or give a command is a very natural, fast, and simple way to get answers to questions, open apps, control your smart home, and much more.

After all, talking to your tech gadget is more intuitive than typing or tapping, and getting a humanlike response is more meaningful too. So Apple Watch wearers will no doubt benefit from the fact that the watch has a built-in microphone and speaker.

An Internet connection is required to send your words to Apple's servers for processing. So as long as your watch is on the Internet via Wi-Fi or wirelessly tethered by Bluetooth to your iPhone, or if you own a cellular-supported model, Siri might just be the best feature of your Apple Watch. You might not always be in a place where you can talk openly (such as in a quiet boardroom meeting), or you might not have Internet access at that moment (such as on an airplane without Wi-Fi or cellular support), but most of the time, you can use Siri to get what you want — quickly.

But you don't know where to start, you say? No problem. I cover Siri extensively in this chapter. In fact, there is some iPhone stuff in this Chapter, often because it's related to what your Apple Watch can do with your personal assistant — so bear with me.

Shameless plug alert: As the author of *Siri For Dummies* (John Wiley & Sons, Inc.), I show you in this chapter all the ways you can use Siri to get information on your Apple Watch.

Setting Up Siri on Your Apple Watch

Before we get going, be aware you have four ways to call up your personal assistant on Apple Watch:

>> Press and hold the Digital Crown button.

>> Say "Hey Siri."

>> Raise your hand toward your mouth.

>> Select the Siri watch face (see Chapter 4 for info), add the Siri complication, and then tap it to talk to Siri.

First, the good news: You probably set up Siri on your iPhone when you first turned on your device, so you have access to your assistant on Apple Watch, too. As you may or may not recall from previous chapters, your iPhone asked you whether you wanted to enable Siri (and, yes, you can enable or disable it on your iPhone by tapping Settings ➪ Siri & Search). Figure 7-1 gives you a look at the Siri options on the iPhone.

Selecting a language

When you set up your iPhone, it also asked for your language preference. You have more than 40 options, including English, French, Spanish, Italian, German, Chinese, Japanese, Korean, and Russian. The reason you choose a language and dialect isn't just so Siri can speak in a language you understand; it's also to let your new personal assistant understand you better. People in the United States and Canada say "Call Mom" differently from English-speaking people in the United Kingdom and Australia, for example, where that phrase may sound more like "Coll mum" or "Cull mam." In fact, you can choose nine versions of English for Siri: Australia, Canada, India, Ireland, New Zealand, Singapore, South Africa, United Kingdom, and United States Obviously, Americans have various accents too — differences definitely exist among speakers from Long Island, Boston, Dallas, and Minneapolis, for example — but American English can be vastly different from the English spoken in London or Sydney. Be sure to choose the correct version, or you may have some difficulty understanding Siri — and vice versa.

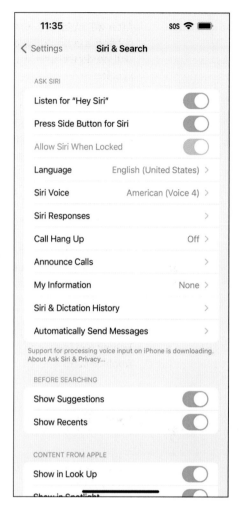

FIGURE 7-1:
Select your Siri options on your iPhone, such as voice gender and language.

It's important to note that Siri has a female voice in the United States by default, but you can change the voice to male if you like. For this reason, I usually refer to Siri as "it" to keep the language universal. Also, in the spring of 2021, Apple introduced new male and female voices to Siri, including two African-American options, to offer more diversity. You can hear and select the new options on your iPhone by tapping Settings ⇨ Siri & Search ⇨ Siri Voice.

If you change the voice for Siri on iPhone, it will be the same voice on Apple Watch. Got it?

To make sure that Siri works flawlessly on Apple Watch, first tap Settings ⇨ Siri & Search on your iPhone, and flick the toggle switch to green (on) for the following three features:

>> **Listen for "Hey Siri."** You can decide whether you want to say "Hey Siri" instead of pressing the Home button on an iPhone (an older model that has a physical Home button) or pressing the Digital Crown button on Apple Watch. If you enable this feature on your iPhone, you can simply raise your wrist to your mouth and say "Hey Siri," followed by your question or command.

   Note: This feature must be activated on iPhone in order for "Hey Siri" to work on your Apple Watch! See below.

>> **Press the Side Button for Siri (iPhone X and later).** Make sure that this option is turned on (toggle switch is flicked to green). This option allows you to press the Digital Crown button on your Apple Watch to initiate your personal assistant. (On iPhone 8 and older models and iPhone SE, you need to press and hold the Home button to activate Siri.)

>> **Allow Siri When Locked:** Yes, hopefully you lock your iPhone with a PIN or Face ID. Otherwise, someone could access your information on a lost or stolen device, which isn't good. Enabling this feature means you can activate your personal assistant without having to unlock the phone first. (And no, if someone found your phone, they couldn't ask Siri to unlock it!)

On Apple Watch, your options are similar, as you'll see in the Watch app on iPhone (Settings ⇨ Siri):

>> **Listen for Hey Siri:** Your personal assistant will answer your queries through the watch if a nearby iPhone isn't detected. If it is, you'll hear Siri through your phone. You must enable this feature on iPhone for your Apple Watch to listen for "Hey Siri."

>> **Raise to Speak:** If you like the idea of raising your wrist to wake up your personal assistant on your Apple Watch, tap Settings ⇨ General ⇨ Siri, and turn Raise to Speak on. You can also enable or disable the "Hey Siri" verbal option here instead of on your iPhone.

>> **Press Digital Crown:** If you prefer, press and hold the Digital Crown for a moment and your personal assistant will wake up and be ready to answer your questions or commands.

>> **Siri Responses:** By default, you'll hear Siri's voice through your Apple Watch speakers, but you can change it to silent or headphones only. This can be toggled in Settings ⇨ Siri & Search on your iPhone or Settings ⇨ Siri on your Apple Watch.

You can activate Siri on your Apple Watch or iPhone and tell it to call you something else. Instead of going by Robert, you can say "Siri, call me Bob" or "Siri, call me Junior." Going forward, Siri addresses you by the name you prefer. You can always change it if you like.

TIP

As you may know, sometimes Siri just doesn't say things right. This is especially true for some names and places — perhaps with origins in other languages — that may be difficult for Siri to pronounce. And you can't blame Siri if it's spelled one way but pronounced another. (For example, I have a friend named Alissa, but it's pronounced "Aleesa.") If Siri says something wrong, just tell her. After she mispronounces something, say "That's not how you pronounce *XXX*." Siri will ask for the correct pronunciation and let you check she got it right!

Ready to rock?: Connecting and talking to Siri on iPhone

Before you get started, you need to ensure that you have a good Internet connection. As I cover in this chapter, you need to be online for Siri to work. You can see the strength of Wi-Fi and cellular signals on your Apple Watch; without a good connection, you might find Siri to be inaccessible.

Okay. To connect to Siri, follow these steps:

1. **Make sure that your Apple Watch is connected to the Internet via cellular, Wi-Fi, or Bluetooth (and/or connected to your nearby iPhone).**

2. **On your paired iPhone, tap Settings ⇨ Siri & Search.**

    Make sure that the Listen for "Hey Siri" option is on. On iPhone X or later, make sure that the Press Side Button for Siri option is on. On iPhone 8 or earlier, make sure that the Press Home for Siri option is on.

3. **On your Apple Watch, tap Settings ⇨ Siri.**

    Choose whether you want to turn "Hey Siri" and Raise to Speak on or off. (See the previous section for more information on these options.)

Talking to Siri on Apple Watch

Before you start talking, know that you can maximize Siri's performance on Apple Watch by following these tips:

>> **Speak clearly.** I know your own speech can be difficult to be conscious of, but the less you mumble and the more you articulate your words, the better Siri

works. Don't worry: Siri is remarkably keen on picking up what you say (and even what you mean), so you don't need to speak like a robot. Just be aware that you'll get better results with clearer speech.

>> **Find a quiet place.** A lot of background noise isn't great for Siri because it may not be able to pick up what you're saying very well. The quieter the environment is, the better Siri can understand your instructions. Finding a quiet place may be tough if you're dining in a crowded restaurant, driving with the window open, or walking down a busy street, of course, so you may need to speak a little louder and closer to the Apple Watch microphone.

When you ask Siri a question — such as "What's the weather like in Seattle tomorrow?" — you should see a ball of colors swirl around the bottom of the watch screen to confirm that it's listening to you. Stop talking when you're done, and you should hear a beep to confirm that Siri is processing your request.

If you make a mistake while asking Siri a question (maybe you accidentally said the wrong person's name to text), or if Siri didn't hear you clearly, you can tap the screen to nullify the request and then ask again. You should hear the familiar ping tone to confirm that Siri is listening for your new request.

The final thing you should see is Siri performing your desired action. Siri might open a map, an email message, a calendar entry, or a restaurant listing, or show you information such as the score of your favorite team's last game (without even opening an app). Depending on what you ask Siri, you may see the information rather than hear it. If you're after a dictionary definition or a numerical equation, for example, you might hear something like "Here you go" or "This might answer your question" before Siri shows you the information on the screen. Other times, Siri both tells and shows you the answer.

TECHNICAL STUFF

Because all requests to Siri are uploaded to a server, it's not unheard of for the server to be temporarily inaccessible, but that situation doesn't happen very often. Siri will apologize to you and ask that you please try again later. A problem with Siri *isn't* an indication of a problem with your Apple Watch or iPhone, so don't fret. Outages usually last only a couple of minutes (if that long), but you should be aware that they can occur.

What Are Siri Shortcuts?

In iOS 12 and later, Siri Shortcuts let you perform everyday tasks with the apps you use most by simply asking for Siri or tapping your Apple Watch or iPhone. You can choose whether this feature works on your iPhone's Lock screen too. See Chapter 4 for Shortcuts tips and tricks.

When it comes to Siri and Shortcuts, your personal assistant learns your routines across your favorite apps and then suggests an easy way to perform common tasks. If you typically ask for weather at the same time of day, for example, Siri may prompt you with the info from your favorite weather app. If you like to order a coffee to pick up every morning with the same app, Siri might suggest the beverage that you tend to pick. Now, that's smart!

Siri makes suggestions for what you might want to do next, such as call into a meeting or confirm an appointment, based on your routines and how you use your apps.

A few more examples:

>> When you start adding people to an email or calendar event, Siri suggests the people you included in previous emails or events.

>> If you get an incoming call from an unknown number, Siri lets you know who might be calling — based on phone numbers included in your emails.

>> If your calendar event includes a location, Siri assesses traffic conditions and notifies you when to leave.

>> When using the Safari web browser, Siri suggests websites and other information in the search field as you type.

>> Above the keyboard, Siri also suggests words and phrases based on what you were just reading.

>> As Siri learns which topics you're interested in, they'll be suggested in the News app.

You can also add Shortcuts to Siri. Look for the Add to Siri button in your favorite apps — hundreds are already supported — and then tap to add with your own personal phrase. Or go to Settings to find all shortcuts available on your device.

Using the Siri watch face and Siri Shortcuts

As I cover in Chapter 4, you can choose many watch faces for your Apple Watch. The aptly named Siri watch face is tied to your personal assistant; updates throughout the day; and shows relevant content and information that you may need based on your location, time of day, and routines. You might see calendar events, boarding passes, or your favorites from the Home app (including controlling and monitoring your smart home devices). The Siri watch face also supports Siri Shortcuts. Figure 7-2 shows Siri in action.

To use the Siri watch face, follow these steps:

1. **Press and hold the watch face on your Apple Watch.**

2. **Swipe left, and tap the New (+) tab.**

3. **Swipe down or twist the Digital Crown button until you land on the Siri watch face.**

4. **Tap the center of the screen to select that watch face.**

5. **Select Add Face (near the bottom).**

Using Siri effectively on Apple Watch

Everything you can do with your fingers, you can do with Siri — if not more, and in less time. A good way to demonstrate its versatility is to look at a few built-in Apple Watch apps and some examples of how you can use Siri to get what you want.

Siri works on the main Home screen, while accessing the clock, or in any app you find yourself in. For example, you can ask Siri for map directions even though you're using the Music app at the time.

TIP

If you're unsure about all the things Siri is capable of doing, say "Hey Siri, what can you do?" on your iPhone or Apple Watch. You should see a huge list of things!

Clock/World Clock apps

Examples of using Siri for time-related tasks include:

» "What time is it?"

» "What time is it in Dubai?" (Figure 7-3 shows what you might see after asking this question.)

» "What time will the sun rise in Brisbane?"

» "How many days until Christmas?"

» "What day of the week will it be on August 15, 2024?"

FIGURE 7-3:
Ask all kinds of time-related questions, and you'll get answers, such as the local time or (as shown here) the time in another city.

It's 5:18 PM in Dubai, the United Arab Emirates.

Dubai, the United Arab Emirates
Current Time

5:18 P.M.
+8HRS

Messages app

Examples of using Siri to send and receive messages include:

>> "Read me my messages."

>> "Do I have any messages from [name]?"

>> "Text my spouse 'Hey, hon, how's your day going?'"

>> "Text 212-555-1212 'I'm looking forward to our after-work drink tonight.'"

>> "Text Julie and Frank 'Where are you guys?'"

>> "Reply 'That's awesome news!'"

TIP

Something fun to try with Siri on Apple Watch — which may amuse the kids — is to ask Siri to remind you of something really far in the future. I asked Siri to remind me to kiss my wife in 10,000 years, and Siri asked whether that event should be placed in my calendar then. (I hope someone finds a cure for mortality soon!)

Phone/Contacts apps

Examples of using Siri to make calls or look up Contacts information include:

>> "Call Mom."

>> "Dial 212-555-1212."

>> "What's Michael Smith's address?"

>> "What's my sister's work address?"

>> "Learn how to pronounce my name."

>> "Show Jennifer's location."

Mail app

Examples of using Siri for looking for email include:

>> "Show me my email" or "Check email."

>> "Do I have any email from David Smith?" (Figure 7-4 illustrates what happens when you ask for email from a specific person or company. Swipe down to see the correspondence between you and this person.)

>> "Show the email from Natasha yesterday."

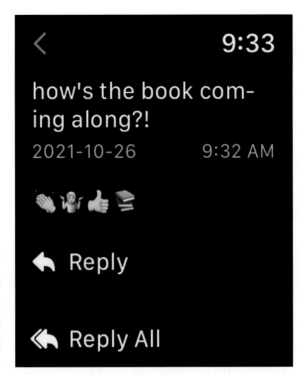

FIGURE 7-4:
Ask for email,
and Siri shows
it to you,
whether you
want to view
it by time
or person/
company.

Calendar app

Examples of using Siri to access calendar information include:

> ❯❯ "Show me what appointments I have on Monday."

> ❯❯ "When is my next meeting?"

> ❯❯ "When's my next appointment?"

> ❯❯ "Move my 12 p.m. meeting to 1 p.m."

> ❯❯ "Cancel the meeting at 4 p.m."

Activity/Workout apps

Examples of using Siri for fitness-related tasks include:

> ❯❯ "Open the Activity app."

> ❯❯ "Open the Workout app."

> ❯❯ "See Move information in Activity."

» "See Stand information in Activity."

» "See Exercise information in Activity."

» "Open Indoor Walk in Workout app."

» "What's my heart rate?"

Maps app

Examples of using Siri to look for directions or a local business include:

» "Show my location on a map."

» "Where is the closest coffee shop?"

» "Take me home." (Figure 7-5 shows a sample result. Siri might ask you where you live the first time you say this, or it will pull the information from your Contacts page.)

» "Take me to Grand Central Station."

» "What's my next turn?"

» "Give me directions to Mom's office."

» "Find a gas station."

» "Find the best sushi restaurant in Miami."

Music app

Examples of using Siri to play music include:

» "Play Harry Styles." (Figure 7-6 shows another example.)

» "Play Workout playlist."

» "What song is this?"

» "Shuffle my music."

» "Play rock music."

» "Play 'Super Freaky Girl' by Nicki Minaj."

» "Skip this track."

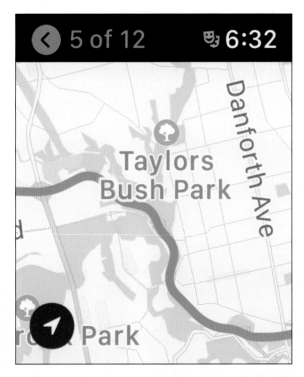

FIGURE 7-5:
Use your voice
in the Maps
app — perhaps
to take you
home.

FIGURE 7-6:
Ask Siri to play
music by artist,
song, album,
genre, and
playlist. The
song could
be streaming
from a service
such as Apple
Music (left) or
stored on a
nearby iPhone
(right).

Speaking of music, in late 2019, Apple Watch gained the capability to Shazam a song — that is, use the Shazam app to find out the name of a playing track and who sings it — even over cellular connectivity if you don't have your iPhone nearby. As shown in Figure 7-7, when you hear a song that catches your ear, you can raise your wrist and ask what it is.

FIGURE 7-7: You can ask Siri to identify a song for you with the help of Shazam.

Web searches

It's easier than ever to use Siri for a web search on your Apple Watch. The results appear onscreen for you to read and scroll through, and you can tap a URL to go to the appropriate page. As shown in Figure 7-8, Siri is more helpful than ever before, including scouring web pages for you.

Miscellaneous

Other examples of random but fun things you can do with Siri on Apple Watch include:

>> "What's the weather outside?"

>> "What's it going to be like this week?"

>> "Do I need a coat?"

>> "How are my stocks doing?"

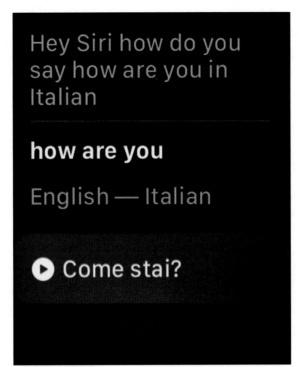

FIGURE 7-8:
Ask, and ye
shall receive!
Apple Watch
displays web
results tied to
Siri queries.

>> "What's the Apple stock price?"

>> "How's the Dow Jones doing?"

>> "Set an alarm for 7 a.m."

>> "Set an alarm for one hour from now."

>> "Wake me up in 30 minutes."

>> "Open Passbook."

>> "Open Stopwatch."

>> "Open Timer."

>> "Open Settings."

WARNING

Because it takes only a quick Siri request to set up a reminder, you might be tempted to do so while driving. But even a minor distraction could cause an accident, so resist using Apple Watch and Siri until you've parked the car.

Trying Other Tasks with Siri

Siri is one of the fastest, easiest, and most accurate ways to interact with content on your Apple Watch. Siri can perform lesser-known yet impressive feats too, and the following sections discuss some of my favorites. Just start your request with "Hey Siri," and Siri will ask how it can assist you.

Setting reminders by location

It's a breeze to ask Siri to remind you of something by time, such as "Tomorrow at 10 a.m., remind me to call the dentist to book an appointment." But did you know that you can set reminders by location too?

You might raise your wrist and say, "Hey Siri, remind me to call Mom when I leave here." Whenever you leave wherever you are — such as your office, a coffee shop, or a shopping mall — Siri will remind you to call your mom. Your nearby iPhone's integrated Global Positioning System (GPS) feature means that Siri is location-aware.

Another example: Say "Remind me to take out the trash when I get home." Because Siri accesses your home address from your iPhone's Contacts app, you won't be reminded of the chore until you pull into the driveway.

If you haven't added your home address information yet, see "Setting Up Siri on Your Apple Watch" earlier in this chapter for directions.

Reading your texts

Many Siri users are aware that they can dictate text messages. Simply say something like "Hey Siri, text Mary Smith 'Please don't forget to call the florist for tomorrow's event.'"

But did you know that you can have your text messages read to you? Raise your wrist or press the Digital Crown button and then say something like "Read my texts." After Siri reads a message to you, you can say something like "Reply saying 'That's an excellent idea — thanks'" or "Tell her I'll be there in 20 minutes."

You can also ask Siri something like "Do I have any texts from Mary?"

Calculating numbers

Siri gets support from Wolfram|Alpha's vast database of facts, definitions, and even pop-culture information. (Ask Siri who shot J.R. or Mr. Burns!) But you can also ask Siri to perform math problems for you. If you're adding up checks to deposit at the bank, for example, ask Siri something like "What's $140.40 plus $245.12 plus $742.30 plus $472.90?" Within a moment, you should hear the correct answer (which is $1,600.72).

If you're out with friends at a restaurant, and the bill comes to, say, $200, you can also ask Siri something like "What's an 18 percent tip on $200?" Siri tells you how much that is ($36).

Siri can also handle multiplication, subtraction, equations, fractions, and more. All you have to do is ask!

Finding your friends

If you're not familiar with the Find People app on Apple Watch or the Find My app on the iPhone, these apps use GPS to provide your geographical location to the people with whom you choose to share this information, such as your spouse, kids, grandkids, friends, or co-workers. When you add consensual people, you can also see their whereabouts on a map — represented by colored orbs — and get the addresses if you so desire.

You probably saw this one coming: You can use Siri to get the most from the Find People app. Raise your wrist and say, "Hey Siri, where are my friends?" The Find People app opens, and you should see who's around and how far they are from you. Now you can send someone a message such as "Let's grab a coffee" on your phone. You can also make queries like these:

>> "Is my husband at home?"

>> "Where's John Smith?"

>> "Find my sister."

>> Where's Julie?"

To get started, go to the Find My app on your iPhone to set things up on your Apple Watch and to add friends. (You can't do it on Apple Watch.) Then you can enjoy this feature because all models (except the original 2015 Apple Watch) have integrated GPS.

And remember, you can also use your voice to find compatible items, too (see Chapter 6).

Extending the Fun (and Silly) Ways to Interact with Siri

Siri is pretty funny, if you haven't yet figured this out from talking to it on your iPhone. In case you haven't, the following sections describe some fun and cheeky things you can ask Siri on your Apple Watch — and the kinds of responses you can expect.

Spoiler alert: Read only the questions, not the answers, if you want to see how Siri replies on your own!

Say "What's the best smartwatch?"

You don't expect Siri to recommend a rival Android-powered watch, do you? Instead, it answers this question with something like "The Apple Watch will show you a really good time" or "I say Apple Watch — hands down."

Say "I love you, Siri."

Deep down, Siri might be flattered, but it suggests otherwise. Siri might say something like "You hardly know me" or "That's nice — can we get back to work now?" or "Impossible!"

Say "Siri, I'm bored."

If you find yourself bored while wearing your Apple Watch, you can tell Siri how you're feeling, and it may reply with something like "Not with me, I hope." Or it'll converse with you, offering a story, song lyrics, or a poem, or engaging in an exchange of "knock, knock" jokes if it's in the mood.

Say "Who's your daddy?"

This one borders on the naughty. Although Siri may be a little reluctant to answer at first, it knows which side its bread is buttered on. You might hear "You are" or perhaps something like "I know this must mean something — everybody keeps asking me this question."

Say "What's the meaning of life?"

You can ask Siri a profound question, such as "What's the meaning of life?" it might give you a literal translation or a cheeky reply such as "A movie" or "All evidence to date suggests it's chocolate." Or it might answer "I don't know, but I think there's an app for that."

Say "Will you marry me?"

After professing my affection for Siri several times (it responded "That's sweet," "I sure have received a lot of marriage proposals lately," and "You are the wind beneath my wings"), I went for it and asked for Siri's . . . uh, hand in marriage. Its reply: "Let's just be friends, okay?"

Chapter **8**

Apple Watch as Your Workout Buddy and Digital Doctor

F itness is one of the smartest applications on your smartwatch. Whether you're trying to monitor your regular daily activity or lose weight, whether you're an athlete looking to maximize your training or a nonathlete who wants to manage workout regimes in an easy way, Apple Watch can handle it all. While you wear one of these high-tech — yet water- and sweat-resistant — devices on your wrist, you'll receive real-time information, such as distance traveled, calories burned, and many other details. Consider it a pedometer on steroids! Apple Watch also serves as a digital doctor of sorts, offering a bevy of health-related applications on your wrist. And although Apple Watch can show your activity information on its small screen, you can dive deeper on your iPhone, which can track more details and historical information because it syncs with the same apps.

REMEMBER

The first-generation Apple Watch (2015) doesn't support fitness- and health-related features. Be sure to have the latest operating system installed (watchOS 9, at this writing) to take advantage of the latest features.

Tracking Your Fitness and Health with Apple Watch

Apple has really evolved its Apple Watch to become an indispensable device for health and fitness. A combination of sensors (hardware), apps (software), and services (like Apple Fitness+) can really help keep you in top-top shape. Here's a look at what this little wrist-mounted gadget can do:

>> **Accelerometer:** Like other activity trackers, Apple Watch has a built-in accelerometer that can count your number of steps, like an old-fashioned pedometer.

>> **Activity trends:** Apple Watch can send activity trends data to your iPhone, which then stores daily trend data for active calories, exercise minutes, stand hours, stand minutes, walk distance, and cardio fitness. Trends compares your last 90 days of activity to the last 365.

>> **Altimeter/barometer sensor and GPS:** The built-in altimeter calculates the number of stairs you climb, and in the latest model, Apple Watch Series 8, the altimeter is always on. Its integrated GPS chip tracks how far you've moved. For the Apple Watch Ultra model, you also get precision dual-frequency GPS for even more accuracy, as well as a redesigned compass app with waypoints and backtrack support.

>> **Mindfulness app:** For the mindful, Apple Watch includes an app that makes you take a quick break from your day to focus on your breathing.

>> **Heart-rate sensor:** This integrated feature tracks your workout's intensity and, if used every day, might detect when something is off. It records unusually high or low heart rates and alerts you about them even when you don't feel symptoms. Features include high and low heart rate notifications, irregular rhythm notifications, and cardio fitness notifications.

- » **Electrocardiogram (ECG or EKG):** This feature records the timing and strength of the electrical signals that make the heart beat. By looking at this data, a physician can gain insights about your heart rhythm and look for irregularities. **_Note:_** This feature is unavailable in Apple Watch SE.

- » **Blood-oxygen monitor:** This Apple Watch app measures your blood-oxygen level periodically throughout the day (if background measurements are turned on). A healthy reading typically is between 95 and 100 percent. You can also take an on-demand reading at any time. **_Note:_** This feature is unavailable in Apple Watch SE.

- » **Fall detection:** Beginning with Apple Watch Series 4, Apple Watch includes an accelerometer and gyroscope to detect when you've fallen. You can initiate a call to emergency services, or if you're unresponsive after 60 seconds, the watch automatically places an emergency call and sends your location to your emergency contacts.

- » **Crash detection:** Available in all three 2022 models — Apple Watch Series 8, Apple Watch SE, and Apple Watch Ultra — your Apple Watch can detect a car crash, dial emergency services and your emergency contacts, and share your location information.

- » **Emergency SOS, Siren:** This feature lets you call emergency services. You can also notify your emergency contacts, send your current location, and display your Medical ID badge onscreen for emergency personnel. Apple Watch Ultra also has an 86-decibel siren to attract attention when in an emergency situation (audible up to 180 meters).

- » **Noise app:** Apple Watch can help protect your ears by monitoring nearby noise levels and your duration of exposure.

- » **Cycle tracking:** Designed to help women manage their monthly menstrual cycles, this discreet, easy-to-use app shows relevant cycle information (including retrospective ovulation estimates for family-planning purposes).

- » **Temperature-sensing:** Apple Watch Series 8 and Apple Watch Ultra also have two temperature sensors — one on the back crystal, near your skin, and another just under the display — to sample your temperature while you sleep (every five seconds). This dual-temp sensor design improves accuracy by reducing bias from the outside environment.

This chapter covers all these topics and more. But I start by showing you how to use two of Apple Watch's main fitness apps: Activity and Workout.

Getting Up and Running with the Activity App

As Apple explains on its website, fitness isn't "just about running, biking, or hitting the gym. It's also about being active throughout the day."

Thus, one of the two main fitness apps on Apple Watch is devoted to your general activity levels during a regular day, includes such activities as walking the dog, chasing after your kids or grandkids, and taking the stairs rather than an escalator or elevator. The aptly named Activity app's main screen (shown in Figure 8-1) keeps track of everything physical you do and encourages you to keep moving.

FIGURE 8-1: The Activity app displays multicolored rings based on your movement, exercise, and more.

Quite simply, the Activity app gives you a visual snapshot of your daily activity. It's broken into three colored rings:

>> **Move:** The reddish-pink ring shows how many calories you've burned by moving.

>> **Exercise:** The lime-green ring shows the minutes of brisk activity you've completed that day.

>> **Stand:** The baby-blue ring gives you a visual indication of how often you stood up after sitting or reclining.

Your goal is to complete each ring each day by reaching the suggested amount of exercise, as outlined in this chapter. The more solid each ring is, the better you're doing.

The first (main) screen of the Activity app gives you a Move, Exercise, and Stand summary, but if you swipe to the left, you can access a dedicated screen for each of the Activity meters.

As shown in Figure 8-2, you should see a summary of each Activity section, which explains what you're seeing in this app.

Before you begin any activity, however, Apple Watch wants to know a little about you first — namely, your gender, age, height, and weight — for the numbers to be accurate, such as estimating your calories burned. (A 25-year-old female burns calories at a different rate from a 65-year-old male, for example.) You have to provide this information only once, and you can answer the required questions with your fingertip (see Figure 8-3).

If you live in the United States, you should see customary unit measurements, such as pounds, but those who live in Canada or the United Kingdom fill out information by using the metric system. (The same applies to distances: miles or kilometers.)

You'll also be asked to select how active you are — Lightly (310 Calories/Day), Moderately (590 Calories/Day), or Highly (860 Calories/Day) —and you can tweak all these numbers, too.

Regardless of which Activity screen you're in — the summary page, Move, Exercise, or Stand — you always see a clock in the top-right corner, so you always know the current time without having to leave the Activity app. Smart, no?

Move

Moving is good, even if you're not moving fast. Motion helps you burn calories; it gets your heart pumping and your blood flowing. The Activity app's Move ring tells you how well you're doing based on your personal active calorie-burn goal for the day, as shown in Figure 8-4. In this example, the default goal is 450 calories per day, which is a couple hours of walking around a shopping mall. If that goal is too easy or too ambitious, you can easily make necessary adjustments to suit your needs. Scroll to the bottom of the main Activity screen (with the three rings), and tap Change Goals. Just like you did when you first set up your Activity rings, you can always change the goal to something more achievable by pressing + or − until you see your desired goal. Then tap Next to make adjustments in your Exercise and Stand goals.

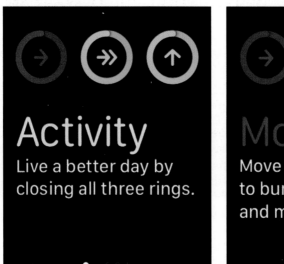
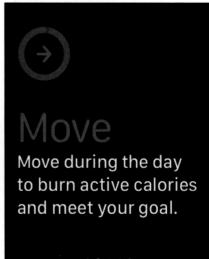

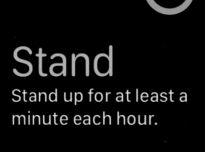

FIGURE 8-2: Apple Watch explains how Move, Exercise, and Stand work. You can also share your Activity info via the iPhone app, which might motivate you to do better!

Activity
Live a better day by closing all three rings.

Move
Move during the day to burn active calories and meet your goal.

Stand
Stand up for at least a minute each hour.

Exercise
Try to get in activity that's at or above a brisk walk.

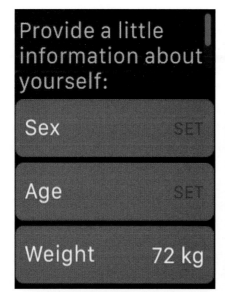

FIGURE 8-3:
Start by answering some questions about yourself.

FIGURE 8-4:
This ring summary screen shows your Move, Exercise, and Stand stats for the day. Just swipe up from the rings screen.

To use the Move tab of the Activity app, follow these steps:

1. **Press the Digital Crown button to go to the Home screen.**

2. **Tap the Activity icon.**

   You can also raise your wrist and say, "Hey Siri, Activity." Either action launches the Activity app. You see the rings that summarize your fitness goals for the day.

3. **Swipe up on the Activity app's main (summary) screen to see additional info.**

   Move tells you how much you've moved during the day. The red number at the top of the screen is your current estimated calories burned and how far you are toward your daily goal (such as 240/320 CAL). You also see your results as a percentage of your daily goal, such as 70%.

4. **Change your caloric goal in the Activity app by scrolling down the main Activity screen, selecting Change Goals, pressing + or – to set your desired goal, and tapping OK.**

   You can change your Exercise and Stand goals in the same fashion (see Figure 8-5).

5. **Swipe up on the screen to see the History graph showing how well you've done for each hour of the day (highlighted by a vertical line).**

   The taller the pinkish bar is, the more you moved that hour, as shown on the left side of Figure 8-6.

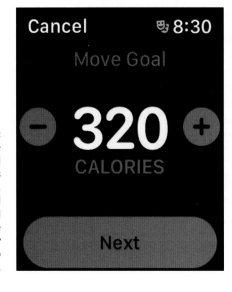

FIGURE 8-5: If you don't like the suggested (default) goals for Move, Exercise, and Stand, scroll down the main Activity screen and tap Change Goals.

6. **Swipe up again or twist the Digital Crown button to see even more details, such as total steps, total distance, and flights climbed.**

   You can see these statistics on the right side of Figure 8-6.

Now you can go about your business.

FIGURE 8-6:
Scroll up within the Activity app to see your day's performance by the hour.

Exercise

Whether you want to do something active in one shot — such as jogging on the treadmill after work — or a little bit here and there, you should try to do at least 30 minutes of exercise each day. What constitutes "exercise," you ask, and how is it different from mere "moving"? Any activity at the level of a brisk walk or above is considered to be exercise, Apple says. By seeing how much you're exercising — or not exercising — you may just be motivated to improve your overall health (which has also been proved to be linked to happiness).

To use the Exercise tab of the Activity app, follow these steps:

1. **Press the Digital Crown button to go to the Home screen.**

2. **Tap the Activity app.**

   You can also raise your wrist and say "Hey Siri, Activity." Either action opens the Activity app, taking you to the main Activity screen.

3. **To go to the Exercise area, swipe down until you see the green text that shows you the percentage of completed exercise goals for the day and how many minutes you've accomplished.**

   Alternatively, at any time, and regardless of the app you're in, tell Siri "Show me Exercise information" to go right to this screen.

4. **Change your goals if you want to, as described in "Move" earlier in this chapter.**

5. **Swipe up to see your History graph, which shows your hourly activity level — measured in minutes — when you were most active.**

   As you might expect, the higher the line goes on the graph, the better. Even if you exercise a little here and a little there, every bit helps and counts toward your daily time goal.

6. **Swipe up again or twist the Digital Crown button to get additional exercise information.**

   You see a numerical summary of your day's achievements. Alternatively, you can grab your iPhone and open the Activity app.

7. **Press the Digital Crown button to return to the Home screen.**

   Don't be discouraged if you're not reaching your exercise goals. Try again tomorrow, or reduce the number of your suggested active minutes — from, say, 30 minutes to 20 minutes.

Stand

Many of us — including yours truly — have jobs that require us to sit for a good chunk of the day. This fact isn't doing much for those love handles we're trying to get rid of. Apple Watch knows when you stand and move around at least for one minute, and all this movement counts toward your Stand ring. Even if you just got up from your computer to get a glass of water, do a small stretch, or walk down the hall to say hello to a co-worker, all this Stand time adds up. You've completed the default Stand ring requirements if you move at least 1 minute in 12 different hours during the day. The app reminds you to get up if you've been idle too long (about an hour).

If you want to disable these reminders, open the Apple Watch app on your iPhone, tap My Watch ➪ Activity, and toggle the Stand Reminder switch to the off (gray) position. Your iPhone syncs with Apple Watch and stops reminding you to get up.

To use the Stand tab of the Activity app, follow these steps:

1. **Press the Digital Crown button to go to the Home screen.**

2. **Tap the Activity app.**

   You can also raise your wrist and say "Hey Siri, Activity." Either action opens the Activity app, taking you to the main Activity screen. The blue ring represents your daily Stand info.

3. **Swipe up to see blue info near the middle of the screen: how many hours you've stood up for (at least one minute per hour), your goal hours (such as 12), and your goal percentage completed.**

4. **Change your goals if you want to, as described in "Move" earlier in this chapter.**

5. **Swipe up to see the History graph.**

   You should see the day laid out chronologically and a full vertical bar for any hour you stood (for at least a minute per hour).

   It doesn't matter whether your time is in consecutive hours or spread out throughout the day; the idea is to get up at least once per hour during the day (unless you sleepwalk, which means that you won't have to worry about your Stand goal overnight!).

6. **Swipe up again or twist the Digital Crown button to obtain more information about your activity.**

   You can get a numeric summary of your day's progress — across Stand, Exercise, and Move. (See the "Understanding the Workout App" section for more.)

Be proud! You're getting your move on.

Understanding the Workout App

You may be wondering how the Workout app differs from the Activity app. Aren't they the same thing?

Not exactly.

Whereas both apps are fitness related, Workout differs from Activity in one respect: Instead of showing your progress over the past day, Workout provides real-time information about calories burned, elapsed time, distance, speed, and pace for your walks, jogs, runs, cycling, and indoor equipment, such as an elliptical, a stair stepper, a rower, a treadmill, and more. In other words, rather than generic daily stats, the Workout app shows you cardio information based on what you're doing while you're doing it.

Although most activity trackers and smartwatches can spit out generic information on your estimated calories burned simply by moving, Apple's technology is tailored to specific exercise equipment and/or exercises. All you have to do is choose the type of workout you'd like to tackle; Apple Watch turns on the appropriate sensors, such as accelerometer and gyroscope (for motion), heart-rate monitor, and GPS (if exercising outdoors, for example, because GPS requires a line of sight with satellites above the Earth).

Then you can receive a detailed summary of your exercise — and, of course, your workout counts toward your Activity ring measurements for the day. You can also set goals, chart your progress, and earn awards.

Not including the first-generation Apple Watch, the Workout app can track swimming too. Don't try this with other fitness trackers or smartwatches unless you know for certain that they're waterproof!

To use the Workout app on your Apple Watch, follow these steps:

1. **Press the Digital Crown button to go to the Home screen.**

2. **Tap the Workout app.**

   You can also raise your wrist and say "Hey Siri, Workout." Either action launches the Workout app's main screen. You should see options for many kinds of indoor and outdoor exercises, as shown in Table 8-1.

   Most of the exercise options are self-explanatory. The exceptions are the following:

   • *High-intensity interval training (HIIT):* This type of exercise alternates intense movement with rest. You might jump rope for 45 seconds, rest for 30 seconds, and repeat with another exercise.

   • *Other:* For this option, the app recognizes that you might not find a matching workout type for your activity. You still earn the calorie or kilojoule equivalent of a brisk walk any time sensor readings are unavailable.

3. **Swipe up or down to view a workout, or cycle through workouts by twisting the Digital Crown button forward or backward.**

   What are you in the mood for? Figure 8-7 shows some of your options. You have many more than are listed here!

TABLE 8-1

Workout App Exercise Options

Outdoor Walk	High Intensity Interval Training
Outdoor Run	Hiking
Outdoor Cycle	Rower
Indoor Walk	Stair Stepper
Indoor Run	Yoga
Indoor Cycle	Functional Strength Training
Dance	Wheelchair
Elliptical	Core Training
Cooldown	Open Water Swim
Pool Swim	Tai Chi
Pilates	Multisport Triathalon

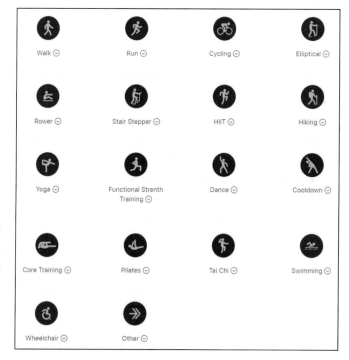

FIGURE 8-7: Choose a workout and tap it to set a goal. Some categories having both indoor and outdoor measurements.

4. **When you see a workout you like, such as Outdoor Cycle, tap the three little dots in the top-right corner to see and set goals.**

    On this screen, you can select a goal:

    - Calories (highlighted in pink)

    - Time (highlighted in yellow)

    - Distance (highlighted in blue)

    - Open (returns to workout type; highlighted in green)

5. **Press + or – to increase or decrease the goal numbers, respectively.**

    For calories, your goal might be 300; if you choose a time-based goal, you might select 45 minutes; a distance-based goal might be 2 miles. You should see different options based on your activity. If you're running indoors, the watch uses the accelerometer, but cycling outdoors uses the GPS on your iPhone to calculate distance. Make sense?

    Depending on the workout you're in, you may also see Set Pace Alert at the bottom of the screen. If you tap this option before you start your workout, Apple Watch alerts you when you're ahead or behind a set pace after 1 mile.

6. **Tap Open to start the workout.**

    Alternatively, tap the workout you want, such as Outdoor Cycle, and you see a countdown timer (3-2-1) to start the workout you chose.

7. **Do your thing.**

    The watch tracks your every move (see Figure 8-8). Well, *almost* every move. Apple Watch always gives you proper credit for things like push-ups, pull-ups, and crunches. Sure, these things add to your Move tab within the Activity app, but the watch may not properly calculate calories burned in the Workout app. Do them anyway, because you know that they help your health — even if your watch doesn't!

8. **Look at your screen for real-time info.**

    Your Apple Watch screen shows relevant real-time information on your workout (as shown in the left image of Figure 8-9). For Indoor Walking, for example, you see elapsed time, estimated calories burned, speed of average mile, heart rate (beats per minute), and elevation.

    During your workout, you should see progress updates to help motivate you. You should also receive timely encouragement at certain times — maybe halfway through your workout or when you reach a certain milestone, such as 1 mile of a 3-mile jog. Figure 8-9 shows an example.

FIGURE 8-8:
Workout
numbers start
at zero so you
can select
goals. Or don't
set a goal and
go for a run;
Apple Watch
still calculates
your steps,
time, distance,
and calories.

9. **If you need to pause or end your workout, swipe to the right on the screen.**

   You see the options Pause, Stop, Lock (to prevent you from accidentally hitting your screen during a workout, which can pause or stop the analysis), and New (to start a new session). Tap the relevant button, all shown in the right image of Figure 8-9.

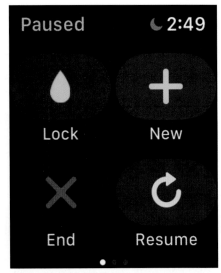

FIGURE 8-9:
Receiving an
alert upon
achieving
a goal (left)
can give you
incentive to
keep going.
You can
always press
the screen
to pause
or cancel a
workout (right).

10. **To play music while exercising, swipe to the right in the Workout screen and press Play, or skip forward and back between tracks.**

Apple Watch remembers how it played music the last time, such as using Apple Music, (see Figure 8-10). See Chapter 9 for more on this topic.

FIGURE 8-10:
Access your
music while
working out
with a simple
swipe to the
left (left). You
may need to
tell the watch
where to
find music to
play, such as
a streaming
service or
locally stored
tunes (right).

11. At the end of your workout, press Stop to view a summary.

After you press Stop, you see a summary of your workout, including total calories burned, active calories burned (when you were physically exerting yourself), resting calories burned, average pace per mile, average heart rate, total distance, and total time of workout. Figure 8-11 shows a few summary screens for a short walk.

To remind you, all the numbers are color-coded, such as total distance in blue, total time in yellow, and active calories burned in pink. At the bottom of the summary screen, you can choose to save or discard this information; tap the option you prefer.

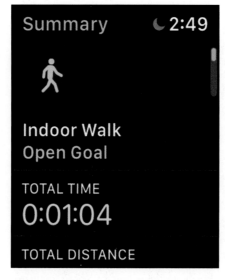

FIGURE 8-11: Swipe to get a report on your workout session (left), including distance, calories burned, and heart rate. You're notified and congratulated if you hit your goal (right).

Now you know how to select, start, and stop a workout, as well as read your summary information. After you review your accomplishments, you can repeat the exercise (with the same goals) to see how you fare, or you might decide to increase or decrease the goal or to choose a different goal. If you started with a time or distance goal, for example, you might change things up by setting a goal based on caloric burn.

Introducing New Workout Views, More Customization, Heart Rate Zones, and More

The introduction of watchOS 9 in 2022 added a number of new features in the Workout app to explore. For one, the Workout display now lets you get more information while working out. (See Figure 8-12.) Just twist the Digital Crown for new views of metrics like Activity Rings, Heart Rate Zones, Power, and Elevation.

FIGURE 8-12:
A glimpse at the new Workout view options while exercising.

If it's an Outdoor Run or Cycle workout you do often, you can now choose to race against your last or best result ("ghost run") and receive in-the-moment updates, including dynamic pacing data, to help you stay in the know (and motivated). You can also see Stride Length, Ground Contact Time, and Vertical Oscillation, to help understand how efficiently you run. (See Figure 8-13.)

In fact, depending on the type of workout, the metrics you can see on your Apple Watch include:

>> Heart rate

>> Power (an instantaneous measurement of your runs to help you stay at a level you can sustain)

>> Segments

>> Splits

FIGURE 8-13:
The Workout
app provides
instant
feedback about
your run.

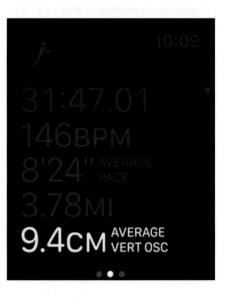

» Elevation

» Pace

» Cadence

» Distance

» Duration

» Vertical oscillation

» Running stride length

» Ground contact time

Speaking of the Workout app, triathletes can take advantage of automatic transitions between swimming, biking, and running. Swimmers can have a new stroke detected, Kickboard (for which you hold onto a foam-based floatation device to swim), plus you can track your SWOLF score for each set (obtained by adding together the number of strokes taken in the pool length and the time it took to swim that length). For example, 40 seconds and 10 strokes to swim the length of a pool will give a SWOLF score of 50.

Personalizing Reminders, Alerts, Feedback, and Achievements

Despite what you've heard, *information* is bliss — not ignorance.

Apple Watch can not only calculate your workouts for you, but also present the data in an accessible way so you can see how well (or poorly) you're doing in reaching your fitness goals. Actually, the watch goes one step further: It can nudge you to be more active, provide weekly goal summaries, and reward you for a job well done.

Reminders

Apple Watch delivers customizable coaching reminders that can help you reach your Activity goals: Move, Exercise, and Stand. You can disable these goals in the Apple Watch app on the iPhone by tapping My Watch ⇨ Activity (left side of Figure 8-14). You can also see a quick-glance summary screen in the Apple Watch app (middle of Figure 8-14), as well as deeper fitness information in the Health app (right side of Figure 8-14).

FIGURE 8-14: Get nudged into activity by setting reminders to shake your booty (left). View how many times you've hit your Activity goals per month or in each category, or get in-depth summaries in the Health app.

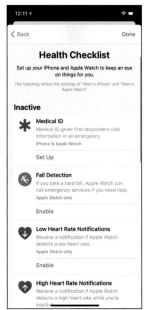

Along with notifying you to get up when you've been idle for too long, over time, Apple Watch learns your goals and accomplishments, and suggests an achievable daily Move goal. You can adjust your fitness goals — bumping them up or trimming them down — to something more reasonable based on your capabilities or time.

Summary

Every Monday, you should receive a weekly check-in, which serves as a summary of your Activity progress. This check-in might say something like this: "Last week's Active Calorie burn goal was 300. You hit it 4 out of 7 days." On the graph that accompanies the text, you see which days you reached your goal and by how much; the higher the vertical column is, the better you did. You should see statistics for all seven days of the week.

You can also open the Activity app to glimpse how you're doing per day, and you can always swipe down from within the Activity app to see an hour-by-hour account of your daily Move, Exercise, and Stand goals. See "Getting Up and Running with the Activity App" and "Using the Fitness App on Your iPhone" in this chapter for more information on these apps.

Beginning with watchOS 9 is a deeper workout summary, allowing you to scrub through a visualization of your run in order to glean more data along the way. (See Figure 8-15.)

FIGURE 8-15: Apple provides more tools to analyze your workouts in the Apple Watch app on iPhone.

Achievements

Isn't it enough incentive to know you're doing well? Well, not always. Let's face it: It's nice to be acknowledged for your efforts, even rewarded.

"Get a pat on the back, right on the wrist," says Apple on its website.

I like that.

Earn special badges, such as the ones shown in Figure 8-16 — although they're a little hard to see because I haven't completed them, so they're not colored. These badges are stored in the Fitness app on your iPhone, which you can look at with pride.

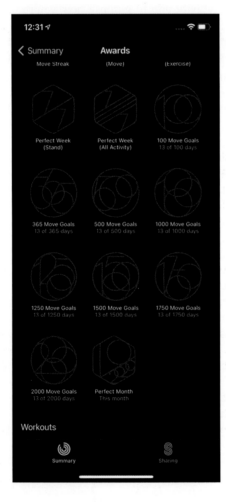

FIGURE 8-16: Apple Watch badges you can earn for reaching a milestone. Most of these badges aren't filled out yet (author hangs his head in shame); otherwise, the award emblems would be colored.

Example Move badges include:

>> **Perfect Month:** Earn this award when you reach your Move goal every day of a single month, from the first day to the last.

>> **Move Goal 200%:** Earn this award every time you double your daily Move goal.

>> **100 Move Goals:** Earn this award when you reach your daily Move goal 100 times.

More important, perhaps badges will encourage you to keep going.

Using the Fitness App on Your iPhone

Only so much information fits on the small Apple Watch screen. Thus, the Fitness app on your iPhone takes things a step further by providing a ton of data based on your Move, Exercise, and Stand achievements, as well as your workout information.

Yes, it might be confusing because this same app (represented by the same multi-colored icon) is called Activity on your Apple Watch but Fitness on your iPhone. In this section, I correctly refer to the app as Fitness (iPhone), which is in sync with your Activity app (Apple Watch).

Okay, back to business.

Specifically, the iPhone app's History tab maps your progress over long periods; you can look at how well you did by day, week, month, or year. Figure 8-17 shows the Activity app on the iPhone.

The Fitness app also displays your badges (of honor!), as well as an Achievements tab with various objectives to unlock as added incentives. Like many video games that reward you for achieving goals — such as "Complete an end-level boss fight in under five minutes" or "Capture a town without losing a life" — your Fitness app can give you fitness-related challenges to take on. By default, the app has a dozen challenges, each represented by a badge-like icon, which can be "Reach your Move, Exercise, and Stand goals in one day" or "Exercise a total of 20 miles."

What's more, the Fitness app syncs its data with your iPhone's Health app, where it can be accessed by third-party health and fitness apps — with your permission, of course.

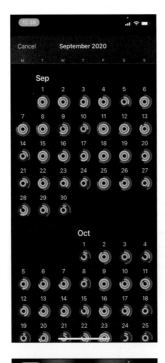
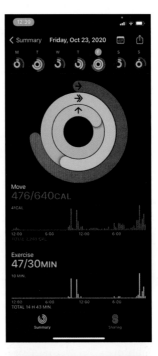
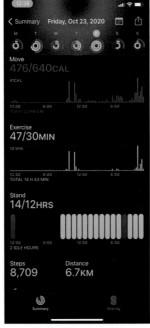

FIGURE 8-17:
The iPhone
Fitness app,
providing
a weekly
summary of
your Activity
levels.

Checking Activity Trends

Since 2019, Apple Watch users have been able to take advantage of the Trends feature, which gives you a deeper look at your activity (or inactivity), over a period of time, which is viewable on your iPhone. Specifically, you can analyze and compare your current Activity rings — Move, Stand, and Exercise — with past performances. That is, you can see a Weekly Summary on your Apple Watch (by scrolling down within the Activity app to the bottom), but for a longer Trends look, it's all in the iPhone's Activity app.

In the Health app on the iPhone, you can see a snapshot of other fitness metrics, including average walking pace, flights of stairs climbed, maximal oxygen uptake levels, and whether you're improving or falling behind. Trends compares your latest 90-day averages with performance over the past 365 days. All this information is laid out in convenient table and graph form (see Figure 8-18), and you even get customized coaching suggestions to get you back on track!

FIGURE 8-18:
A look at the Activity Trends data gleaned from Apple Watch and sent to iPhone's Health app.

Using (and Loving) the Mindfulness App

 Life can be hectic, so Apple's Mindfulness app reminds you to take a moment out of your day to relax and . . . well, breathe.

You're always breathing, of course (if you weren't, you wouldn't be alive to read this book, silly!), but this wrist-based reminder guides you through a series of deep breaths to calm you down. The Mindfulness app (formerly called Breathe) lets you choose how often and how long you want to breathe; then the Apple Watch animation and gentle taps help you focus throughout this short exercise. The app also includes a new session type, Reflect, which welcomes you with what Apple calls a "unique, thoughtful notion to consider that invites a positive frame of mind."

Starting a Mindfulness session: Breathe

Had a rough day at the office? Sat near a crying baby on an airplane? Here's how to initiate a breathing session to calm you down, via your Apple Watch:

1. **Press the Digital Crown button to go to the Home screen, and tap the Mindfulness app to open it.**

2. **Select Breathe.**

3. **Tap Continue and follow along with on-screen instructions to breathe.**

4. **If you want to change the session's length to something other than 1 minute (default), tap the three little dots by the word Breathe and make the changes.**

   Tap where it says 1 Minute and select 2, 3, 4, or 5 Minutes instead.

   When you're done, tap Duration in the top left to go back to the main screen (see Figure 8-19).

5. **Inhale as the animation grows on the Apple Watch screen.**

   You also feel little taps on your wrist.

6. **Exhale as the animation shrinks and the taps stop.**

7. **Breathe until the session ends and your watch taps you twice and chimes (unless Silent mode is on).**

   When you're done, you see your resting heart rate.

FIGURE 8-19:
A look at
Breathe,
part of the
Mindfulness
app, on Apple
Watch. Follow
along with
the animation
and haptic
feedback.

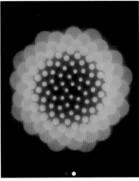

REMEMBER

Because you're supposed to be calm and to focus on your breathing, your Apple Watch mutes some notifications. If you answer a call or text, or move around too much during a session, the session ends automatically, and you won't get any credit for your efforts.

Using the Apple Watch app on iPhone to make Mindfulness changes

You can use the Apple Watch app on the iPhone to make additional Mindfulness settings (left side of Figure 8-20).

1. **Open the Apple Watch app on your iPhone.**

2. **Tap My Watch ⇨ Mindfulness.**

3. **Scroll to near the bottom of the screen, and Breath Rate.**

   You can change the length of each breath — the default setting is seven breaths per minute — and tweak how many Breathe reminders you want per day, how forceful the haptic (vibration) feedback is, and so on.

4. **Check your progress.**

   To check how often you take time to use the Mindfulness app on Apple Watch, track your sessions with the Health app on iPhone (right side of Figure 8-20). Open the app, tap Health Data ⇨ Mindfulness, and then tap the graph.

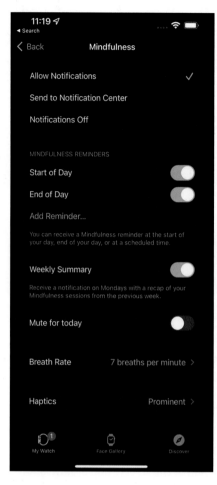

FIGURE 8-20:
The Apple
Watch app on
the iPhone
is where
you make
changes in the
Mindfulness
app (left);
the Health
app on the
iPhone shows
a summary of
your breathing
exercises
(right).

Starting a Mindfulness session: Reflect

With the move from watchOS 7 to 8 in 2021, Apple renamed the Breathe app Mind-fulness. Although breathing is still a big part of the exercises, the app now has a Reflect section as well (see Figure 8-21), offering you something positive to help you focus and center your thoughts, such as "Think of something you did that was fun. Remember why you liked it" or "Recall a time recently when you felt a sense of calm. Bring that feeling into this moment."

While you're reflecting on whatever the watch prompts you to reflect on, you'll see swirling colors as a tranquil background on the screen (or close your eyes, if you prefer).

FIGURE 8-21:
As part of the new Mindfulness app, the Reflect sessions asks you to reflect on something positive to put you in the right frame of mind.

To get going, follow these steps:

1. **Open the Mindfulness app on Apple Watch.**

2. **Tap Reflect, read the theme, focus your attention, and tap Begin.**

3. **Set the duration of a session, such as 1 minute (see Figure 8-22).**

   You can choose a time between 1 and 5 minutes.

Adjusting Mindfulness settings

As you can with all other apps on Apple Watch, you can tweak the Mindfulness app. You can change how frequently you get Mindfulness reminders, for example, or mute the reminders during a workday, change your breath rate or how the haptic (vibration) settings work on your wrist, and more.

To make changes, open the Settings app on your Apple Watch (which looks like a little gray gear), tap Mindfulness, and then do any of the following:

>> **Set and create reminders.** In the Reminders section, turn Start of Day and End of Day on or off; tap Add Reminder to create additional reminders.

>> **Get or stop a weekly summary.** Turn Weekly Summary on or off.

>> **Mute reminders.** Turn on Mute for Today.

>> **Change your breathing rate.** Tap Breath Rate to change the number of breaths per minute.

FIGURE 8-22:
The Reflect
session lets
Apple Watch
users focus
on something
positive for 1
to 5 minutes.
Scroll through
the provided
tips to get the
most out of
these mental
exercises.

>> **Choose haptics settings.** Tap Haptics; then choose None, Minimal, or Prominent.

>> **Get new meditations.** Turn on Add New Meditations to Watch to download new meditations when your Apple Watch is connected to power. Meditations you've completed are deleted automatically.

By the way, you can also open the Apple Watch app on your iPhone, tap My Watch ⇨ Breathe, and then adjust a setting.

REMEMBER

You can see your heart rate during Mindfulness sessions. Yep, complete a Reflect or Breathe session, and your heart rate appears on the Summary screen. You can also review your heart rate later, on your iPhone. Tap the Health app's icon on your iPhone, tap Browse ⇨ Heart ⇨ Heart Rate ⇨ Show More, swipe up, and then tap Breathe.

Finally, as you'll see in Figure 8-23, you can add the Breathe watch face to your Apple Watch to get quick access to Mindfulness sessions.

To change your watch face to Breathe, follow these steps:

1. **Press and hold your Apple Watch's current watch face.**

2. **Swipe left all the way to the end, and tap the New button (+).**

3. **Turn the Digital Crown button to select Breathe.**

4. **Exit out to the home screen by pressing the Digital Crown again.**

FIGURE 8-23: Some people may like using the Breathe watch face to get quick access to Mindfulness sessions.

Enabling the Flashlight feature on Apple Watch

This chapter is all about fitness and health — and that includes safety as well, in my book. A little later in the chapter, I discuss crash detection, for example, but there's a little-known flashlight feature I wanted to share with you.

Say you're walking to your car in a dark parking lot and need a little bit of extra light. Your Apple Watch has a handy flashlight you can enable to illuminate your walk.

No, it's not an actual light that beams out of the Apple Watch, but rather a white screen that covers the entire face, and there are some modes within it, too, like a strobe.

Here's how to enable it:

1. Swipe up from the bottom of the Apple Watch display to open Control Center, and then tap the Flashlight icon on the left. (See Figure 8-24.) Alternatively, tell Siri to "open Flashlight"!

2. When you see the bright screen, angle your wrist so that it brightens up your walk or whatever you're trying to see in the dark.

3. Within the Flashlight app, swipe left or right for a flashing "strobe light" effect and a red light, too, which some people may prefer.

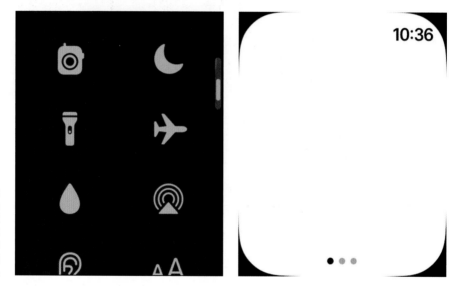

FIGURE 8-24: Using the Flashlight feature on Apple Watch to illuminate dark environments.

Hearing Health: Using the Noise App

Enjoy hard rock concerts? Work with heavy machinery? Apple Watch's Noise app can show you how loud ambient audio is near you and how long you've been exposed to it.

Using the Noise app is quite straightforward. Follow these steps:

1. Tap the Noise app's icon.

2. Enable the microphone, which gives the app permission to listen to the environment.

The app doesn't record or save the audio. It shows you the estimated noise level around you, measured in decibels (the higher the number is, the more the sound can damage your hearing), as shown in Figure 8-25. The number is complemented with a color-coded system for convenience. If the reading is green, you're good; if it's yellow, turn down the sound. You can tap the reading to get more info.

TIP

You can also add the Noise app to your favorite watch face as a complication. See Chapter 4 to find out how.

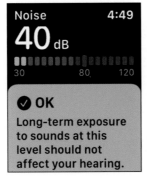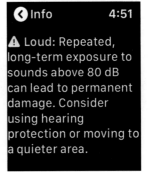

FIGURE 8-25: As Quiet Riot once sang, "Cum on feel the noize!" It's easy with the Noise app on Apple Watch.

Advanced Health Help: Heart-Rate, ECG, and Blood Oxygen Monitoring

Apple Watch can do a lot more than just provide activity information and workout tracking. Although you now know that it measures your resting heart rate and how hard your heart works while you're exercising, you may not know that it can alert you if something seems to be a little off. You also may not know that since Apple Watch Series 4, your wrist-mounted device has had an electrocardiogram (ECG or EKG, depending on the source) feature, fall detection, and SOS features.

Before I get into these features, I'm sure that some of the more technology-minded readers will want to know how Apple Watch does all these things. So this section starts by explaining the workings of each monitor in all its glory.

How the heart-rate sensor works

Underneath the Apple Watch case — the part that touches your skin — is a ceramic cover with sapphire lenses. Below that part is a series of small Taptic

Engine sensors, vibrate slightly to tell you something, such as when you have a message from someone. In that same area, you have a heart-rate monitor that logs the intensity of your workout or your resting heart rate.

TIP

If someone you know also has an Apple Watch, you can send them your heartbeat. See Chapter 5 for more about how to do this, as well as send taps and sketches to other Apple Watch owners.

The heart-rate sensor uses infrared light, visible-light LEDs, and photodiodes to detect your heart rate, measured in beats per minute (BPM). The average heart rate is 72 BPM, but when you start your workout and your muscles need more oxygen, your heart beats faster to pump oxygen-filled blood throughout your body.

Every 10 minutes, Apple Watch measures your heart rate and stores that information in the iPhone's Health app. This data, along with other information the watch collects about your movement, goes toward estimating the number of calories you're burning.

The Apple website explains that the heart-rate sensor uses *photoplethysmography*: "This technology, while difficult to pronounce, is based on a very simple fact: Blood is red because it reflects red light and absorbs green light." Apple Watch uses green LED lights, paired with light-sensitive photodiodes, to detect the amount of blood flowing through your wrist. "When your heart beats, the blood flow in your wrist — and the green light absorption — is greater," Apple says. You can see a map of these amazing sensors in Figure 8-26.

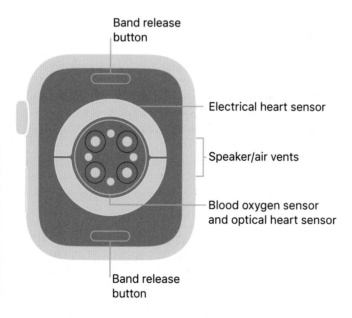

FIGURE 8-26: As illustrated on Apple's website, here's a diagram of the magic behind the Apple Watch Series 6. These sensors are truly amazing, no?

By flashing its LED lights — in an alternating fashion and hundreds of times per second — Apple Watch calculates the number of times the heart beats per minute. Apple says that heart-rate sensor can also use infrared light to measure your heart rate every 10 minutes. But if this infrared system isn't giving a reliable reading, your smartwatch switches to the green LEDs.

For the heart-rate monitor to work effectively, Apple Watch must fit snugly; if your watch band is too loose, the back of the watch case might not touch your skin enough or might move around during a reading. All available Apple Watch bands allow you to tighten them a bit.

Other factors can also affect a reading:

>> **Weather:** Readings might be off if it's too cold out.

>> **Movement:** Fast or constant movement, such as an intense game of squash, can jostle the watch too much for an accurate heart-rate measurement.

>> **You:** Some people just don't give a good reading. Some Apple Watch owners who have major tattoos, for example, report the heart-rate monitor won't work. Why, you ask? The sensor works by shining green light into the wearer's wrist, but it can be confused by the presence of dark tattoos inked under the skin.

If you're not getting a good reading, don't forget that Apple Watch and the iPhone can work with external heart-rate monitors, such as a chest strap, using wireless Bluetooth technology.

How the ECG monitor works

Apple unveiled its super high-tech ECG feature in the fall of 2018, along with its Apple Watch Series 4 models (which you'll need to have to take advantage of this feature). The feature is also available in Apple Watch Series 5 (2019), Apple Watch Series 6 (2020), Apple Watch Series 7 (2021), Apple Watch Series 8, and Apple Watch Ultra (2022), and presumably it will go forward in all flagship models (not in the Apple Watch SE models).

As you might expect from this feature's name, you can use Apple Watch to detect a dangerously high or low heart rate or an irregular heart rhythm (*arrhythmia*).

How does the sensor work? As you know, the back of Apple Watch measures your heart rate on your wrist. To read ECG pulses, touch the Digital Crown button with a finger of your opposite hand; this touch, coupled with pulses captured by the sensors in your watch, creates a closed circuit with your heart. Now your watch can record ECG information and detect when something is off. Figure 8-27 illustrates what happens.

FIGURE 8-27:
Although the technology seems like science fiction, Apple Watch can not only detect and display your heart rate (left), but also serve as an ECG machine that senses abnormalities (right).

Blood-oxygen monitor

If you can't tell by thumbing through this chapter, Apple Watch is slowly evolving into a health and wellness device. One of the most fascinating additions to Apple Watch, beginning with Series 6, is a blood-oxygen sensor that can measure the oxygen level of your blood, anytime and anywhere, providing you valuable insights into your overall wellness.

REMEMBER

That said, Apple always cautions its users *not* to rely on Apple Watch for medical information, because there's always room for error, so blood-oxygen measurements taken with Apple Watch Series 6 or Series 7 are designed only for general fitness and wellness purposes.

Although exceptions and anomalies exist, blood-oxygen levels between 95 and 99 percent are considered to be healthy. A lower number, such as below 80 percent, may lead to compromised respiratory, cardiac, and/or brain functionality, medical experts say. Generally speaking, the higher the percentage of oxygen your red blood cells carry from your lungs to the rest of your body, the better.

Oxygen saturation (SpO2) is a measurement of the amount of oxygen-carrying hemoglobin in the blood relative to the amount of hemoglobin not carrying oxygen. You need to have a certain level of oxygen in your blood; otherwise, your body may not function efficiently.

Low blood-oxygen levels (*hypoxemia*) have been linked to COVID-19, in fact, as the coronavirus may affect your ability to get enough oxygen. Hypoxemia can be caused by other heart and respiratory issues as well, such as pneumonia, asthma, and congenital heart disease. By monitoring your SpO2 levels, Apple Watch may be able to warn you of potential health issues before it's too late.

(Hey, I bet you didn't realize you were going to get a science lesson when you picked up this book.)

How the blood-oxygen monitor works

Sometimes referred to as a *pulse oximeter,* a blood-oxygen monitor — including the one in Apple Watch Series 6 and Series 7 — shines red and green LEDs and infrared light into your wrist. (The back crystal of your Apple Watch houses these lights.) Photodiodes measure the amount of light reflected back. Then advanced algorithms use this data to calculate the color of your blood. The color determines your blood-oxygen level. Bright red blood has more oxygen, and dark red blood has less.

Apple Watch's blood-oxygen monitor works in two ways. One way is passive measurement, in which the watch takes regular readings on you, even when you sleep. Apple calls these readings background measurements. The other way is active measurement; you open the app and request a reading on demand.

What you need to use the blood-oxygen monitor

Before you get going, you need to have and do the following:

>> Make sure that the Blood Oxygen app is available in your country or region. You'll know when you set up Apple Watch, as shown in Figure 8-28. Or check out www. apple.com/watchos/ feature-availability.

>> Ensure that you have an iPhone 6 or later and update it to the latest version of the iOS operating system software (iOS 15 at this writing).

FIGURE 8-28:
Apple Watch's blood-oxygen monitoring discreetly looks out for anything suspicious, or you can get an on-demand reading by launching the app.

>> You need an Apple Watch Series 6 or later running the latest version of the watchOS operating system (watchOS 8 at this writing).

Note the following things:

>> The Blood Oxygen app is not available for use by people younger than 18 years.

>> It's not available if you set up your Apple Watch with Family Setup (see Chapter 2).

Setting up the blood-oxygen monitor for the first time

Perform these simple steps to get going:

1. **On your iPhone, open the Health app.**

2. **Follow the onscreen prompts.**

   If nothing happens, tap Browse ⇨ Respiratory ⇨ Blood Oxygen ⇨ Set Up Blood Oxygen.

3. **Open the Blood Oxygen app on your Apple Watch to measure your blood-oxygen levels.**

Taking a blood-oxygen reading

For an on-demand blood-oxygen reading, follow these steps:

1. **Rest your arm comfortably on a flat surface, such as a table or desk.**

   Make sure that your Apple Watch is snug on your wrist (not too loose).

2. **Tap to open the Blood Oxygen app on your Apple Watch.**

3. **Remain still, tap Start, and keep your arm steady for 15 seconds.**

   The countdown starts, as shown in Figure 8-29.

   When the measurement is done, you receive the result as a percentage (such as 98%).

4. **Tap Done to close.**

   A log of all your readings (with time stamp) is in the Health app on your iPhone (tap Respiratory ⇨ Blood Oxygen).

For a passive blood-oxygen reading (background measurements), the Blood Oxygen app on your Apple Watch occasionally measures your blood-oxygen levels (usually when you're not moving).

TIP

If you want to stop these readings while sleeping or watching a movie — to avoid the distracting light shining from your Apple Watch — open the Settings app on your Apple Watch, tap Blood Oxygen, and turn off In Sleep Mode and In Theater Mode.

Enabling handwashing detection

It took a global pandemic to remind us how important regular handwashing is, and now your technology can help you with that task. As you likely know, washing your hands for 20 seconds can help you stay healthy, as it helps remove viruses, bacteria, and germs you might have picked up from a gas pump, door handle, and so on.

To enable reminders and a timer on Apple Watch (Series 4 and later), follow these steps:

1. **Open the Health app on your iPhone.**

2. **Tap Browse.**

3. **Scroll to the bottom of your screen and tap Other Data.**

4. **Scroll down to the Get More from Health section, and tap Handwashing.**

   You can enable many settings, as shown in Figure 8-30.

Another way to enable reminders is to open the My Watch app on your iPhone and tap Handwashing.

Either way, you're reminded to wash your hands (or you can turn off reminders if you like). Whenever you start handwashing, your Apple Watch automatically detects the motion and counts down from 20 seconds. You feel a little haptic touch, too.

Data on how frequently and long you washed your hands is stored in the Handwashing area of the Health app on the iPhone.

Advanced Health Help: Medical ID, Medications, Fall Detection, and SOS Calling

Wait — this chapter isn't done just yet! Your Apple Watch can do even more to help you remain safe.

Setting up or editing your Medical ID information

Before you delve in to the various Apple Watch health features, a discussion of how to set up your Medical ID information is in order. Why? A lot of these features require this information in the event that something goes wrong with you, and you want to inform a close contact or emergency services.

With that in mind, you need to teach your Apple Watch who is important to contact. You must turn on Wrist Detection for your watch to call emergency services automatically.

Bear in mind falls are also automatically recorded in the Health app on your iPhone unless you reply that you didn't fall when your Apple Watch asks. To check your fall history, open the Health app on your iPhone, tap the Health Data tab, and then tap Results.

To set up your Medical ID, follow these steps:

1. **Open the Settings app on your Apple Watch, tap SOS.**

2. **Tap Medical ID, followed by Set Up. If you have Apple Watch 7 or 8, with the larger screen, you can input data here with the onscreen keyboard.**

    Alternatively, open the Apple Watch app on your iPhone, and tap My Watch ⇨ Emergency SOS.

    Tapping Edit takes you to the Health app on the iPhone. Or simply open the Health app on your iPhone and tap the Medical ID tab.

3. **Tap Edit, and enter your date of birth and other health info.**

    Depending on where you live, you may also be prompted with questions about organ donation (and can sign up, if you like). See Figure 8-31.

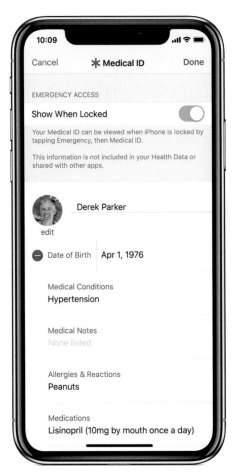

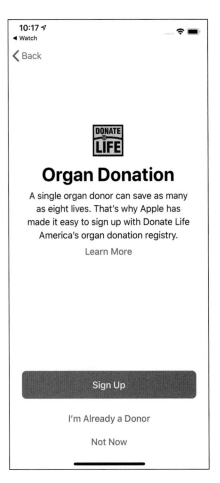

FIGURE 8-31:
Use the iPhone
Health app
to create a
Medical ID that
may be sent
to emergency
responders
via your Apple
Watch (left).
You may also
be prompted
with other
options, such
as donating
your organs
(right).

To add emergency contacts, follow these steps:

1. **Open the Apple Watch app on your iPhone, and tap My Watch ⇨ Emergency SOS.**

   This allows you to have Apple Watch send a message saying that you have called emergency services. Your current location will be included in these messages.

2. **Tap Edit These Contacts in Health to select various people.**

   This opens up people in your Contacts app.

3. **Tap to select emergency contacts.**

   Note: Only mobile phone numbers can receive Emergency SOS alerts.

4. **Tap "Add Emergency Contacts."**

   Figure 8-32 shows the Emergency SOS screen.

5. **Tap the red minus button next to an emergency contact to remove that person and then tap Delete.**

6. **To make your Medical ID available from the lock screen, turn on Show When Locked and then tap Done.**

   In an emergency, this option gives information to people who want to help.

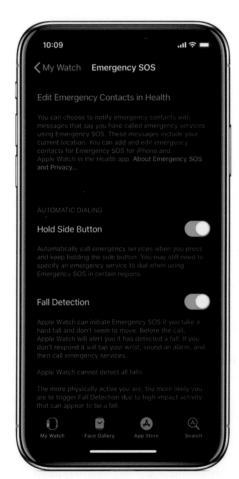

Medications

As we learned about in Chapter 6, Apple Watch is ideal for setting reminders, because it "taps" on your wrist to remind you about something. Added to watchOS 9 in 2022 is specific medication reminders app called Medications, which is super handy. As the name suggests, the app helps you discreetly and conveniently keep track of your medications, vitamins, and supplements, and nudges you when it's time to take them. To get going, use the Health app on iPhone to set up medication logging. See Figure 8-33.

FIGURE 8-33:
Ideal for those who want to keep track of meds, vitamins, and supplements, the new Medications app can help you stay on your game. On the right is the Health app, where you set up your medication details.

To set up Medications:

1. **Open up the Health app on iPhone. Select Browse, and choose Medications.**

2. **In the Medications section, tap Add a Medication, and type the name in the search window.**

3. **Tap Next and answer some questions about the medication, like the kind of medication (tablet, liquid, topical, and so on), strength and unit, and when you take it (date and time).**

   You can even choose the shape and color to help you identify the pill(s) you need at a certain time.

Now, when it's time to take your meds, your Apple Watch's Meds app will buzz on your wrist.

The Medications area of the Health app on iPhone also has a section called **Get More from Health**, on why it's important to track your medications, how to connect to your provider to see past notifications, get notified about updates, and how to add it to your schedule. Tap the blue Get Started tab to fill in information.

Checking your heart rate

At any time, you can check your heart rate with the Heart Rate app. Press the Digital Crown button, open the app, and wait a few seconds for Apple Watch to measure your heart rate.

TIP

From the Heart Rate app on Apple Watch, you can also view your resting, walking, breathing, workout, and recovery rates throughout the day (see Figure 8-34).

FIGURE 8-34: View heart-rate info after you take a reading (left), or look back at different points of the day through the Apple Watch app (middle). The Health app takes a deeper dive into BPM data (right).

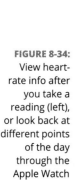

Alternatively, you can summon Siri to show your heart-rate info, and you can add the Heart Rate complication to your watch face.

Receiving heart-rate notifications

Here's something you may not know about Apple Watch: If your heart rate remains above or below a chosen BPM, your Apple Watch can notify you. To get notifications, follow these steps:

1. **On your iPhone, open the Apple Watch app.**

2. **Tap My Watch ⇨ Heart.**

3. **Tap High Heart Rate, and choose a BPM level.**

    If you're unsure what number this should be, consult your physician.

4. **Tap Low Heart Rate and choose a BPM level.**

Now when Apple Watch measures your heart rate after 10 minutes of inactivity, it notifies you if your BPM is higher or lower than the numbers you set in steps 3 and 4 (see Figure 8-35).

FIGURE 8-35: When the Heart Rate feature is activated, you are alerted if your heart rate is too low or too high.

REMEMBER

When you use the Workout app, Apple Watch measures your heart rate continuously during the workout and for three minutes after the workout ends to determine a workout recovery rate.

Turning on Fall Detection

With Apple Watch Series 4, Apple added a new, improved accelerometer and gyroscope (another motion sensor that can detect whether you've fallen). Do you like the idea of this feature? Can you see a loved one benefiting from it? Here's how to turn it on or off:

1. **Open the Apple Watch app on your iPhone.**

2. **Tap My Watch ⇨ Emergency SOS.**

3. **Turn Fall Detection on or off.**

The next section walks you through what to do if you fall.

What to do after a fall

When a fall incident occurs, the watch delivers a hard-fall alert, taps you on the wrist, sounds an alarm, and displays an alert.

REMEMBER

Although handy, Fall Detection isn't perfect. Apple Watch can't always detect a fall if you went down slowly. Conversely, it might misinterpret a fast drop as being a fall when in fact you were doing something physical.

As shown in Figure 8-36, when Fall Detection is activated, you've got a choice to make:

>> **You fell, but you're OK.** You can dismiss the alert by tapping I Fell, but I'm OK.

>> **The call was a false call.** If you didn't fall, tap I Did Not Fall.

>> **You fell, and you're not OK.** If you've just had a bad fall, you can tap and drag the Emergency SOS icon on your watch.

REMEMBER

You must set up your emergency contact list ahead of time, as discussed in "Setting up or editing your Medical ID information" earlier in this chapter.

FIGURE 8-36:
You'll get some options to tap through if you fall. The example on the right shows what happens if you swipe Emergency SOS. You can call 911 or one of your contacts.

If you're unresponsive after 60 seconds, Apple Watch begins a 15-second count-down while tapping you on the wrist and sounding a loud alert (which gets louder so that you or someone nearby can hear it). When the countdown ends, your Apple Watch automatically contacts emergency services. At the same time, it sends a message to your emergency contacts with your geographical location, letting them know that Apple Watch detected a hard fall and dialed emergency services.

REMEMBER

If you entered your age when you set up your Apple Watch (or in the Health app on the iPhone) and indicated that you're 65 or older, the Fall Detection feature turns on automatically.

Using Emergency SOS

Whether it senses a troubling anomaly through the heart-rate or ECG sensor or detects a hard fall, Apple Watch can dial emergency services, notify your emergency contacts, send your current location, and even display your Medical ID badge for emergency personnel. Or you can initiate a call through your watch if you require some emergency attention. Just press and hold the side button, and you'll be connected with emergency services.

Apple Watch now lets you make emergency SOS calls internationally. According to Apple, the feature works almost anywhere for cellular versions of Apple Watch Series 5 and later.

WARNING

In all these scenarios, to make an emergency call, you must have an Apple Watch GPS + Cellular model (with active cellular plan), be connected to an iPhone via Bluetooth, or have access to Wi-Fi (with Wi-Fi Calling set up, as discussed in Chapter 5).

If you start the countdown by accident, simply let go of the side button. Or if you start an emergency call by accident, press the display firmly and then tap End Call. You may get a call back from an emergency-response operator; you can tell them that you dialed the number by mistake.

To call emergency services manually, follow these steps:

1. **Press and hold the side button on your watch until the Emergency SOS slider appears.**

The slider may take two seconds or so to display.

2. **Continue to hold down the side button.**

A short countdown begins, and an alert sounds. Alternatively, you can drag the Emergency SOS slider to the left.

3. **When the countdown ends, your watch automatically calls emergency services.**

4. **When the call ends, your Apple Watch sends your emergency contacts a text message with your current location unless you choose to cancel.**

If Location Services is off, it temporarily turns on.

TIP

If your location changes — such as when a first responder is taking you to the hospital — your contacts get an update, and you too get a notification a few minutes later.

To stop sharing your location, tap Stop Sharing in the notification.

Enabling the siren on Apple Watch Ultra

Only available on Apple Watch Ultra — Apple's largest and most durable Apple Watch, released in late 2022 — an integrated siren lets you quickly and easily alert people around you if you become lost or injured, or if you have any other reason you need help.

Apple Watch Ultra plays a continuous siren at 86 decibels that repeats at regular intervals, which can be heard up to 600 feet away! The siren is a unique and high-pitched sound that includes two distinct patterns not generally heard in nature or the environment that alternate and repeat.

Note: Siren pauses in certain situations, like during phone calls, and when timers or alarms sound, but not while podcasts or music are playing. The siren continues emitting its loud sound until you turn it off (or your watch runs out of battery).

There are two ways to enable the siren:

>> **Action button:** Turn on the siren by pressing and holding the Action button on the left side of the Apple Watch Ultra until the Siren slider appears. Drag the Siren slider to start a countdown. It looks like a red megaphone. After the countdown, the siren starts. See Figure 8-37.

Alternatively, you can press and hold the Action button to start the countdown, then continue holding the Action button until the siren starts.

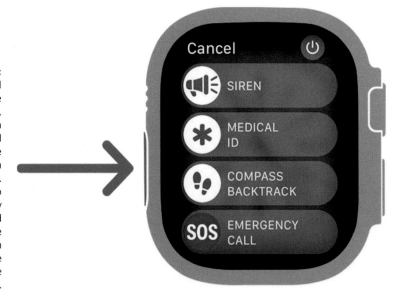

>> **Side button:** You can also press and hold the side button until the Siren slider appears. Drag the Siren slider to start a countdown. After the countdown, the siren starts. Be aware you can cancel the siren during the countdown: If you are holding the Action button, release the Action button to cancel. If you initiated the countdown by dragging the Siren slider, rest the palm of your hand on the watch display for at least three seconds to cancel.

To stop the siren while it's sounding, simply tap Stop in the Siren app.

Understanding Crash Detection on Apple Watch

As the author of *Apple Watch For Dummies*, I gotta tell you: I love this feature.

If your Apple Watch Series 8, Apple Watch SE (2nd Generation), or Apple Watch Ultra detects a severe car crash, it can help connect you to emergency services and notify your emergency contacts.

Currently, the Crash Detection feature is designed to detect severe car crashes — front-impact, side-impact, rear-end collisions, and rollovers — be they in sedans, minivans, SUVs, pickup trucks, or other passenger cars (but not in buses or trains, for example).

On by default, when your Apple Watch detects a severe car crash, it displays an alert on your wrist (and/or on your iPhone 14 and iPhone 14 Pro) and automatically initiates an emergency phone call after 20 seconds unless you cancel. If you are unresponsive, it will play an audio message for emergency services, which informs them that you've been in a severe car crash and gives them your latitudinal and longitudinal coordinates with an approximate search radius. (See Figure 8-38.)

FIGURE 8-38: All three 2022 models of Apple Watch offer crash detection, which is on by default. A potentially life-saving feature, your Apple Watch can contact emergency services with your location information. Wow!

If you've set up your Medical ID, your device displays a Medical ID slider so that emergency responders can access your medical information.

Reminder: Don't forget to set up (and occasionally review) your emergency contacts on iPhone and Apple Watch. Set up your Medical ID and your emergency contacts in the Health app on iPhone. To share your location with your emergency contacts, turn on Location Services for Emergency SOS: On your iPhone, tap Settings ➪ Privacy & Security ➪ Location Services ➪ System Services, and make sure Emergency Calls & SOS is turned on.

To turn crash detection off:

1. **Open the Settings app on your Apple Watch.**
2. **Go to SOS ➪ Crash Detection, then turn off Call After Severe Crash.**

Understanding Cycle Tracking

If you're female, you'll find one of the newer Apple Watch features, Cycle Tracking, to be helpful. You can gain insight into your monthly menstrual cycle to get a clearer picture of your overall health. By logging your cycles, your Apple Watch can predict your next period and fertile window.

Information you can access easily on your wrist can also help track irregularities and symptoms you may want to discuss with your physician.

To use Cycle Tracking, follow these steps:

1. **Open the Health app on your iPhone.**

2. **Tap Cycle Tracking.**

    If you don't see this option, tap Browse in the bottom-right corner of your screen and then tap Cycle Tracking.

3. **Tap Getting Started ⇨ Next and answer the questions onscreen.**

    Questions include "When did your last period start?" (select date), "How long does your period usually last?" (such as 5 days), and "How long is your typical cycle?" (such as 28 days).

4. **Still in the Health app on the iPhone, toggle these options.**

    - *Period Prediction*: Allows the Health app to use the data you entered to predict your period

    - *Period Notifications:* Notifies you of upcoming periods and sends prompts to log them

    Other options include Fertile Window Prediction, Log Fertility, and Log Sexual Activity.

 When you've completed the questions and options in the Health app on the iPhone, you can use the Cycle Tracking app on Apple Watch.

Use the Health app on iPhone to log daily information about your menstrual cycle. You can add flow information (Had Flow, No Flow, or Flow Level [Light, Medium, or Heavy]); record symptoms (such as Abdominal Cramps, Acne, Appetite Changes, Bloating, and Headaches); and see cycle length and variation.

You can even track results from an ovulation prediction kit and readings from a basal body thermometer. All this info is viewable in a colorful graphical chart in the Health app on your iPhone.

Temperature Sensing

Introduced in 2022, Apple Watch Series 8, and Apple Watch Ultra can gather wrist temperature data to help provide insights into your overall well-being.

But this feature is for use while you sleep. That is, your Apple Watch won't tell you if you've got a fever during the day. At least not yet.

Your body temperature fluctuates naturally while you sleep and can vary each night due to things like alcohol consumption, diet, exercise, your sleep environment, or physiological factors (such as illness and menstrual cycles).

Apple says after about five nights, your Apple Watch will determine your baseline wrist temperature and look for nightly changes to it. If you use Cycle Tracking, your wrist temperature can also be used to provide retrospective ovulation estimates — for family-planning purposes — and improve period predictions, too.

To get going, you must enable Track Sleep on your Apple Watch (see below), with Sleep Focus enabled for at least four hours a night for about five nights.

Viewing the history of your nightly temperature is handled on your iPhone, and not the Apple Watch. There is no Temperature app, or anything like that. To view your wrist temperature data, follow these steps:

1. **On your iPhone, open the Health app.**

2. **Tap Browse, then tap Body Measurements.**

3. **Tap Wrist Temperature. (See Figure 8-39.)**

   You'll likely see "Needs More Data" at the top of your chart while your

FIGURE 8-39:
An example of what it's like to view your wrist temperature data in the Health app on iPhone.

baseline temp is still being established. Under your chart appears the number of nights remaining before your wrist temperature data is available. Again, it takes about five nights to re-establish your baseline temperature.

To turn off Wrist Temperature, open the Watch app on iPhone, tap Privacy followed by Turn Off Wrist Temperature.

Along with the two temperature sensors on Apple Watch Series 8 and Apple Watch Ultra — one on the back crystal and another just under the screen — Apple Watch's advanced algorithms then use this data to provide an aggregate for each night that you can view as relative changes from your established baseline temperature on the iPhone's Health app.

TIP

To get the best results, make sure your Apple Watch fits just right. If the fit is too loose, it can impact wrist temperature data.

Tracking Your Sleep with Apple Watch

Now that Apple Watch lasts a long time between charges, many people are wearing it to bed to gain a better understanding of their night's sleep.

The Sleep app is designed to provide that information, so in this section, I explore this handy application in greater depth. The app icon looks like a little white bed against a turquoise background. (Or is it teal?)

When you wake up, tap the Sleep app to find out how much sleep you got and see your sleep trends over the past 14 days (see Figure 8-40).

Note: If your Apple Watch had less than 30 percent battery left before you went to bed, you're prompted to charge it. In the morning, look at the greeting screen to see how much charge remains.

Setting up sleep tracking

To get going, open the Sleep app on your Apple Watch, and follow the onscreen instructions.

You can create multiple schedules, such as one for your Monday-to-Friday workdays and another for weekends. For each schedule, you've got all these options to personalize:

>> A sleep goal (how many hours of sleep you want to get)

>> What time you want to go to bed and wake up

>> An alarm sound to wake you up

>> When to turn on sleep mode, which limits distractions before you go to bed and protects your sleep when you're in bed

>> Sleep tracking, which uses your motion to detect sleep when Apple Watch is in sleep mode and worn to bed

TIP

You can also select all these options by opening the Health app on your iPhone and tapping Browse ⇨ Sleep ⇨ Get Started (in the Set Up Sleep section).

FIGURE 8-40:
On Apple Watch or your iPhone, the Sleep app shows how much sleep you got (because of the sensors in the watch), as well as historical sleep data.

Setting the wake-up alarm

To change or turn off your next wake-up alarm, follow these steps:

1. **Open the Sleep app on your Apple Watch.**

2. **Tap your current bedtime (see Figure 8-41).**

3. **To change your wake-up time, tap the current setting, turn the Digital Crown button to set a new time, and then tap the Check button.**

4. **If you don't want your Apple Watch to wake you in the morning, turn off Alarm.**

FIGURE 8-41:
A look at the Sleep app, which shows the evening's sleep schedule (bedtime and wake up time).

The changes apply only to your next wake-up alarm. After that, your normal schedule resumes.

To exit Sleep mode, turn the Digital Crown button to unlock the watch, swipe up to open Control Center, and tap the Sleep button.

TIP

You can also turn off the next wake-up alarm in the Alarms app, if you prefer. (The app's orange-and-white icon has an old-school clock on it.) Tap the alarm in the Sleep | Wake Up section, and select Skip for Tonight.

Changing your sleep schedule and options, plus Sleep Focus

To change or add a sleep schedule, follow these steps:

1. **Open the Sleep app on your Apple Watch.**

2. **Select Full Schedule.**

3. **Do any of the following:**

   - *Change a sleep schedule.* Tap the current schedule.

   - *Add a sleep schedule.* Tap Add Schedule.

   - *Change your sleep goal.* Tap Sleep Goal and set the amount of time you want to sleep.

   - *Change Wind Down time.* Tap Wind Down and set the amount of time you want sleep mode to be active before bedtime.

   Apple says that sleep mode turns on during Wind Down time, which turns off the watch display and turns on the Do Not Disturb feature.

4. **To complete the change or addition to your sleep schedule, do any of the following:**

   - *Set the days for your schedule.* Tap your schedule, tap below Active On, choose days, and then tap Done.

   - *Adjust your wake time and bedtime.* Tap Wake Up or Bedtime, turn the Digital Crown button to set a new time, and then tap Set.

   - *Set the alarm options.* Turn Alarm off or on, and tap Sound to choose an alarm sound.

   - *Remove or cancel a sleep schedule.* Tap Delete Schedule (at the bottom of the screen) to remove an existing schedule or tap Cancel (at the top of the screen) to cancel creating a new one.

This is also a good time to introduce Focus, a feature introduced to Apple's Apple Watch, iPhone, and more.

Essentially, Focus helps you stay in the moment when you want to concentrate on an activity: Personal, Sleep, and Work. Or you can create a custom Focus on your iPhone (Settings ⇨ Focus), choosing who is allowed to contact you, which apps can send you notifications, and so on.

Focus can reduce distractions — allowing only notifications you want to receive (ones that match your focus) — and lets other people and apps know you're unavailable.

(*Note:* To have your Focus settings shared across all devices where you're signed in with the same Apple ID, open Settings on your iPhone, tap Focus, then turn on Share Across Devices.) With this in mind, to change sleep options, open the Settings app on your Apple Watch and tap Sleep ⇨ Sleep Focus to adjust these settings:

>> **Turn On at Wind Down:** By default, Sleep Focus begins at the Wind Down time you set in the Sleep app. If you prefer to control Sleep Focus manually in Control Center, turn this option off.

>> **Sleep Screen:** Your Apple Watch display and the iPhone lock screen are simplified to reduce distractions.

>> **Show Time:** This option shows the date and time on your iPhone and Apple Watch during sleep mode.

You can turn sleep tracking and charging reminders on or off from this screen. When sleep tracking is enabled, your Apple Watch tracks your sleep and adds sleep data to the Health app on your iPhone. Turn on Charging Reminders to have your Apple Watch remind you to charge your watch before your Wind Down time and notify you when your watch is fully charged. You can choose a different watch face to display when each Focus is active. For example, when Sleep Focus is active, your Apple Watch can display the Simple watch face.

For more about Focus on Apple Watch, visit https://support.apple.com/guide/watch/apd6640937c4/watchos

New with watchOS 9 is the ability to see how much time you spent in REM, core, or deep sleep, as well as when you might have woken up. Plus, the updated Health app on iPhone lets you see metrics like heart rate and respiratory rate (and how things may change throughout the night). (See Figure 8-42.)

Viewing your sleep history

Are your eyes getting heavy? I'm almost done covering all the sleep-tracking features.

To view your recent sleep history, open the Sleep app on your Apple Watch, and scroll down to see the amount of sleep you got the night before and your average over the past 14 days.

If you prefer, see your sleep history on the iPhone by opening the Health app and tapping Browse ⇨ Sleep.

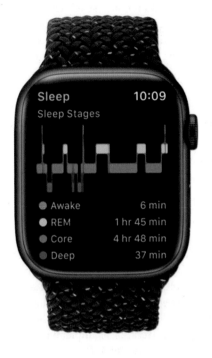

FIGURE 8-42:
We all know
how important
it is to get a
good night's
sleep — and
now your
Apple Watch
can help you
achieve it.

Reviewing your sleeping respiratory rate

Something new and cool has arrived with watchOS 8: You can review your sleep-ing respiratory rate. Yep, your trusty Apple Watch can track your breathing rate as you sleep, which may give you greater insight into your overall health.

After you wake up, follow these two simple steps:

1. **Open the Health app on your iPhone.**

2. **Tap Browse ⇨ Respiratory.**

3. **Tap Respiratory Rate ⇨ Show More Respiratory Rate Data.**

   The Sleep entry shows the range of your respiratory rate as you've slept.

Neat, huh?

WARNING

Apple cautions, however, that respiratory-rate measurements are not intended for medical use.

Using Apple Fitness+ on Apple Watch

Even before the global pandemic hit in early 2020, many people decided to exercise at home instead of going to a gym or fitness studio. Some workouts require equipment, such as a spin bike; others let you exercise with whatever equipment you have in any space you've got available.

In 2020, Apple introduced Apple Fitness+ ($9.99 per month or $79 a year), a video workout service that incorporates the activity metrics captured by Apple Watch for viewing on an iPhone, iPad, or Apple TV device.

You can access studio-style workouts (see Figure 8-43) delivered by inspiring world-class trainers and with motivating music by popular artists. When a workout is selected and started on an iPhone, iPad, or Apple TV, the correct workout type starts automatically on Apple Watch.

FIGURE 8-43: Follow along with trained instructors on one of the many compatible Apple devices, and your Apple Watch will monitor your activity.

At this writing, Apple Fitness+ supports 11 workout types, including: High Intensity Interval Training (HIIT), Yoga, Core, Pilates, Strength, Treadmill (Walk or Run), Cycling, Rowing, Dance, and Mindful Cooldown. Video and audio meditations are also offered as part of the service.

Apple Fitness+ workouts used to require an Apple Watch, but not any longer! You can now sign up and use an iPhone, iPad, or Apple TV — but only Apple Watch captures all the metrics you need for a thorough look at your progress.

During a workout, you can see your heart rate and Activity rings synchronized onscreen. Workouts range from 5 to 45 minutes in length, and new content is added weekly.

If you've completed at least three Fitness+ workouts, your personalized recommendations will appear at the top of the app. If you don't see personalized recommendations, you can still find and filter workouts.

To choose a Fitness+ workout, follow these steps:

1. **On your iPhone, open the Fitness app, and tap the Fitness+ tab.**

    or

    On your iPad or Apple TV, open the Fitness app.

2. **Tap or select a workout type.**

3. **Tap or select the Filter button.**

4. **Filter workouts by type, trainer, or time.**

When you've found a workout, you can do any the following:

>> Tap or select Preview to watch a preview of the workout.

>> Save a workout by tapping the Add button on your iPhone or iPad, or selecting Save Workout on your Apple TV. To access your saved workouts, scroll to My Workouts at the bottom of the Fitness+ screen.

TIP

You can review the music playlist for the workout, but if you have an Apple Music subscription, tap or select Listen in Music to add the playlist in Apple Music.

Here's how you start a workout:

1. **For a given workout in the Fitness app on iPhone, iPad, or Apple TV, tap or select Let's Go.**

2. **Tap or select the Play button.**

    The metrics from your Apple Watch appear onscreen.

3. **Pause and resume your workout as needed on your device or Apple Watch.**

4. **When your workout ends, review your results or start a Mindful Cooldown.**

You can customize which metrics appear onscreen before you start a workout and change them at any time during your workout.

While you're in a workout, tap the Metrics Editor button. (For Apple TV, use your remote to click the Metrics Editor button.) You can display your workout time, heart rate, and calories burned as you work out. Your Activity-rings progress also appears.

You can show how much time has elapsed in your workout or how much time remains, or you can turn off the time display.

TIP

The Burn Bar in Apple Fitness+ is a fun way to compare your effort with that of others who've done the same workout. (Apple says it accounts for weight to allow for equal comparison and is shown as your calories burned.) The Burn Bar is available for the High Intensity Interval Training (HIIT), Treadmill, Cycling, and Rowing workouts.

Chapter **9**

Mucho Media: Managing Your Music, Movies, Apple TV, and More

M usic lovers, listen up: Your Apple Watch can help you get more from your favorite tunes. In this chapter, I cover how to use Apple Watch to listen to music while you're on the go. Although I share how to use your watch to control tracks stored on a nearby iPhone, you don't need your iPhone to listen to music: Your Apple Watch can stream music from the Apple Music streaming service, as well as store and play music on the watch itself.

For the latter two scenarios, there are times when you don't want to take your iPhone with you, but as long as you have Bluetooth headphones (shown in Figure 9-1), you can still enjoy your music.

FIGURE 9-1: You need Bluetooth headphones such as the trendy Apple AirPods to listen to music and podcasts on Apple Watch. Why? The watch has no headphone jack.

I also discuss how to control your music — as well as audiobooks, podcasts, and radio plays — hands-free by using Siri.

Although these capabilities aren't widely publicized, Apple Watch can also control the Apple TV connected to your TV set, your Apple Music app (formerly iTunes) on your Mac, or iTunes on your Windows PC.

 You can share songs, albums, and playlists through Messages and Mail right from the Music app. Tap the Share icon when you're listening to a song and choose how to send it (Messages or Mail) and to whom. See below!

TIP

Using Apple Watch to Control Songs Stored on an iPhone

 Similar to the iPhone app, Apple Watch has a Music app that lets you find and play music stored on your iPhone.

Sure, you can install music on the Apple Watch itself — something I cover in "Syncing and Playing Music from Your Apple Watch" later in this chapter. You can also stream music directly from the Apple Music service (more on this service later), but most wearers will likely use their watch as a kind of wireless remote control.

Sample scenario: You're walking down the street with your swank Beats headphones or wireless AirPods Pro. Rather than take your iPhone out of your jeans pocket every time you want to switch tracks or choose a playlist, simply lift your wrist and perform those functions from the comfort of your Apple Watch. After all, the watch is all about convenient, at-a-glance information when and where you need it. Controlling your music is no different.

To manage your iPhone music from your Apple Watch, follow these steps:

1. **Press the Digital Crown button to go to the Home screen.**

2. **Tap the Music app.**

   You can also raise your wrist and say "Hey Siri, play Music" or something more specific. If you set up Siri to activate the moment you raise your wrist (see Chapter 7), you don't need to summon your personal assistant with your voice at all.

   On the Music screen that appears, you see the name of the song, the artist, and the album artwork (if available), as shown in Figure 9-2. If a song is already playing on your iPhone, you see small equalizer bars dancing at the top of the screen to indicate audio is playing, as well as some scrolling information, such as the complete name of the track, and the album name.

3. **Tap the artwork to play the song, if it isn't already playing.**

   You hear your selection through your iPhone, wired earbuds/headphones, or Bluetooth headset.

   REMEMBER

   You can't hear music through Apple Watch — and wouldn't want to — unless you have a Bluetooth headset or headphones. See "Pairing a Bluetooth Device with Apple Watch" later in this chapter for details on adding one of these devices to your Apple Watch.

4. **Tap + or – to increase or decrease the volume, respectively.**

   A red horizontal bar shows you how loud the volume is getting. You can also skip forward or backward by tapping the double arrows on either side of the Play button. The elapsed time of the track is in the top-left corner, and a digital clock is in the top-right corner.

FIGURE 9-2:
Swipe up
inside the
Music app to
scroll through
album artwork,
and tap to
select a
specific track.

What do you do if you want to hear a new song? You've got many options:

>> Swipe up inside the Music app to display the album art of other songs in your library.

>> Scroll all the way to the top to open a menu with the options Shuffle All, Library, and On Phone (to see music content on your iPhone).

>> Tap the phone or library (which has the music stored on Apple Watch or the Apple Music service) to reveal five options: Artists, Albums, Songs, Playlists, and Downloaded. These options are discussed in detail in the following sections. Another screen is the one you're currently in, called Now Playing.

Now Playing

This screen shows you the artist that's currently playing. You can tap the screen to pause the song or play it again from the beginning. If no album art that Apple Music recognizes has been imported for this track, this area has only text and virtual control buttons, as shown in Figure 9-3.

A dedicated Now Playing app on your Apple Watch Home screen lets you continue playing content on your Apple Watch or iPhone.

Open the Music app on Apple Watch and tap **Library** (for synched or stored songs on Apple Watch) or **On iPhone,** and you'll be presented with these options: Artists, Albums, Songs, Playlists, and Downloaded.

Artists

You can see your music listed alphabetically by artist or band. Swipe up or down with your fingertip or twist the Digital Crown button to review the list. As shown in Figure 9-4, you should see how many songs you have by that person or band.

Albums

As you may expect, this screen is where you can see all your music listed alphabetically by album name. Albums view is ideal when you want to hear an entire album by the same artist, or one or more tracks from an album, or perhaps a compilation or movie soundtrack with multiple artists.

Songs

The Songs screen (see Figure 9-5) is where you can find all your songs listed alphabetically, regardless of artist. Swipe up or down with your fingertip or twist the Digital Crown button to find something to listen to.

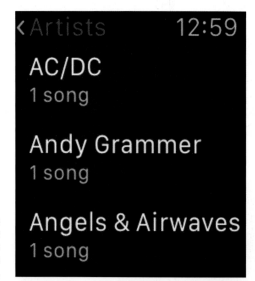

FIGURE 9-4:
Artists view
inside the
Music app on
Apple Watch.

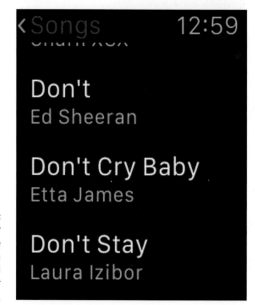

FIGURE 9-5:
Swipe or
twist the
Digital Crown
button to scroll
through your
list of tracks.

TIP

Choosing to see your music collection by song yields the longest list. No worries: You can scroll up or down by using the small bar in the top-right section of the screen. The small bar appears when you turn the Digital Crown button.

Playlists

This screen shows all your music grouped in some fashion, such as by theme, event, or genre (see Figure 9-6). Windows users can create a music playlist in iTunes; Mac users have the Apple Music app. Or create your own playlist on your iPhone or iPad. You might call a playlist something like Driving Tunes, Workout Mix, or Relaxing Music — whatever suits your fancy. If you have a playlist on your iOS device, it's synced with your Apple Watch.

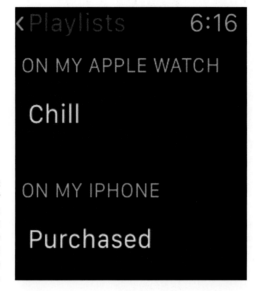

FIGURE 9-6:
Access your custom playlists on your Apple Watch from within the Music app.

In other words, while you can't create a new playlist on your Apple Watch, you can still play it on your watch. You can even store the playlist on the watch if an iPhone isn't nearby.

Downloaded

You can download Apple Music tracks to your iPhone to listen to it when offline — such as when you're without Wi-Fi on an airplane or when you don't want to incur data charges while walking around town — but you can also download and listen to music when you don't have (or want to access) the Internet. Here's how.

REMEMBER

Although you can access a playlist on Apple Watch, you can't create a playlist on it. For that task, you have to sync via the Apple Music app on your iPhone or your Mac, or iTunes on a PC.

You can also sync a playlist to the watch to listen to media when no iPhone is around. See "Syncing and Playing Music from Your Apple Watch" later in this chapter for more on syncing playlists.

You can use Force Touch to display additional options: While listening to a track, press the screen firmly, and you will see options to sync with AirPods (if you own a pair of these wireless headphones) or to Bluetooth the music to a nearby compatible speaker.

Having Siri Play Your Music

Do you know what's even faster than tapping and scrolling to find music? Asking Siri to play it for you. Suppose that you're itching to hear a song in your collection that's been stuck in your head all day. Or maybe you want to give your favorite band's new greatest-hits album a spin from beginning to end. All you need to do is ask Siri (politely) to play an individual song or album. As long as you're in a place where you can talk freely, Siri can launch a song, album, artist, playlist, or genre — whatever you ask for. If you have a cellular model of Apple Watch, you can do everything without your phone!

To have Siri help you play your music via Apple Watch, raise your wrist and then do one of the following three things:

>> Say "Hey Siri, play _____ [name of song, artist, playlist, and so on]."

>> If you enabled Siri to wake up when you lift your wrist (as discussed in Chapter 7), raise your wrist, and ask to play something.

>> Press and hold the Digital Crown button to activate Siri, and ask away.

After Siri processes your request — and don't forget that you need cellular or Wi-Fi access — you should see and hear the music you asked for via your iPhone (or, if the music is stored on your Apple Watch, through Bluetooth headphones).

Here are some examples of what you can ask Siri to do:

>> "Play 'Girls Like You.'"

>> "Play Drake."

>> "Play Road Trip playlist."

- » "Play some hip-hop."

- » "Shuffle my music."

- » "Shuffle The Beatles."

- » "What song is this?" or "Who is this?"

- » "Play similar music."

You can also control your music with your voice. Following are some commands you can give verbally. Press and hold the Digital Crown button (or raise your arm and say "Hey Siri"), followed by

- » "Play" to play the song shown on your watch screen

- » "Pause" to pause the track you're listening to

- » "Skip" or "Next" to go to the next track

- » "Previous song" or "Play previous song" to have Siri play that song

Pairing a Bluetooth Device with Apple Watch

You can load up your watch with music (see "Syncing and Playing Music from Your Apple Watch" later in this chapter) and take it to go, but there's one catch: You can't listen to music through the Apple Watch's tiny speaker (and wouldn't want to). Besides, there's no headphone jack. Instead, you need to pair Bluetooth headphones (or a speaker) with Apple Watch to hear music.

REMEMBER

Pairing isn't necessary if your iPhone is nearby, because you should hear music coming from the iPhone's speakers or from headphones connected to the phone. Pairing a Bluetooth device with Apple Watch directly is required only if your iPhone isn't around.

To connect a Bluetooth device to your Apple Watch, follow these steps:

1. **Put your Bluetooth headphones or speaker in pairing mode.**

   You might see a flashing light on the device to confirm that it's awaiting a connection. The process is different for all products, but pairing usually involves pressing and holding the Home button until you see a flashing light.

2. **On Apple Watch, tap the Settings app's icon.**

   Alternatively, lift your wrist and say "Hey Siri, Settings."

3. **Twist the Digital Crown button until you locate Bluetooth, and then tap that option.**

   Wait a moment, and your Bluetooth headphones or speakers should appear in this list.

4. **Select the Bluetooth device you want to pair it with by tapping its name.**

   Figure 9-7 shows Apple Watch searching for a device.

5. **When your device appears, tap it.**

   You've successfully paired the device! But you still need to change the music source from iPhone to Apple Watch to hear music over your Bluetooth-enabled headphones. Find out how in "Syncing and Playing Music from Your Apple Watch" later in this chapter.

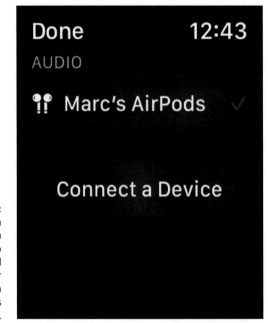

FIGURE 9-7:
Pair a Bluetooth device to hear synced music on your Apple Watch if no iPhone is nearby.

Streaming Apple Music to Your Apple Watch

As you may or may not know, Apple Music is a monthly subscription service from Apple that lets you listen to more than 90 million songs. Unlike iTunes, which lets you pay for and download individual songs or albums, Apple Music is a streaming service, so all the songs are stored on Apple's servers in the cloud for you to access from your device as though they were stored locally. Neat, huh? Instead of owning the music, you're essentially renting the songs, if you will, via your monthly fee.

In other words, Apple Music isn't an a la carte model, but a buffet-style "all-you-can-eat" approach, giving you unlimited access to many songs for a monthly fee. To use this feature, you must have an Internet connection. You have an option to download Apple Music tracks to an iPhone, iPad, Mac, or PC in the event that you know you'll be without the Internet for a while.

Another Apple Music feature is Apple Radio, as we covered earlier in this chapter, which might introduce you to new songs through curated playlists you can stream on demand. You can also use Siri to play any song in the Apple Music catalog and listen to custom stations.

Syncing and Playing Music from Your Apple Watch

Not all smartwatches let you store music on your wrist, so you need a nearby smartphone, but Apple Watch does indeed offer this feature. After all, you don't always have (or want) your phone with you, such as when you go out for a jog for a few minutes. If that's the case, why shouldn't you listen to your favorite high-energy music to keep you pumped up?

This convenience comes with two limitations:

>> **Sound:** As stated earlier, Apple Watch has no headphone jack. You need a pair of Bluetooth headphones paired wirelessly with your smartwatch if you want to hear anything stored on it. See "Pairing a Bluetooth Device with Apple Watch" earlier in this chapter for details.

>> **Space:** Although you have storage for many songs, this space is considerably smaller than what your iPhone, iPad, and computer have. Specifically, you've got up to 2GB of internal storage on Apple Watch dedicated to music and podcasts, which is the equivalent of about 500 songs (at roughly 4 MB each).

If you accept these caveats, follow these steps to transfer music to your Apple Watch from an iPhone:

1. **Connect your Apple Watch to your PC or Mac via its USB charger.**

   Use the special magnetic charger that shipped with your Apple Watch.

2. **On your iPhone, open the Apple Watch app.**

   While you're at it, make sure that Bluetooth is turned on. To double-check, tap Settings ⇨ Bluetooth.

3. **Tap My Watch ⇨ Music.**

   You have to scroll down a bit after the Music screen opens because of its many settings, shown in Figure 9-8, along with apps installed on your Apple Watch.

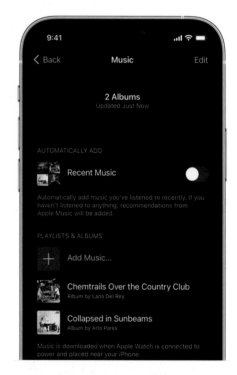

FIGURE 9-8:
Sync music between your iPhone and Apple Watch.

4. **Tap the Heavy Rotation option listed in the Automatically Add section.**

   This option synchronizes the playlists and albums you're listening to with Apple Watch.

5. **To add new albums and playlists, tap the orange + icon next to Add Music.**

6. **Search through music on your iPhone by using the search window or browsing through sections.**

   The sections include Artists, Albums, Genres, and Compilations. You should also see any custom-made playlists you've created on your iPhone, in the Apple Music app on your Mac, in iTunes on a Windows PC (such as Awesome Driving Tunes or Reggae Mix).

TIP

If you can't add specific albums or playlists, download them to your iPhone before you sync with your watch.

7. **Select what you want to transfer.**

Based on which selection you make by tapping the music to sync, the relevant songs are transferred to your Apple Watch. You should see the words **Sync Pending** and then **Synced**. Keep in mind that the more songs you synchronize, the longer it takes to copy them to Apple Watch.

TECHNICAL STUFF

Don't worry about transferring too many tunes over to Apple Watch, because Apple won't let you go over your allotted storage for music: 2GB (about 400 to 450 songs).

Have songs stored on your Apple Watch? To listen to music on your Apple Watch rather than your iPhone, follow these steps:

1. **Open the Music app on your Apple Watch.**

2. **Choose where to play your music from (iPhone, Library, and so on), as shown in Figure 9-9.**

3. **Scroll to Listen Now.**

   You see a curated feed of playlists and albums based on your tastes, as provided by Apple Music.

4. **Add content to your library, create a playlist, and more, as shown in Figure 9-9.**

   If you tap a station, it plays in the Radio app on Apple Watch.

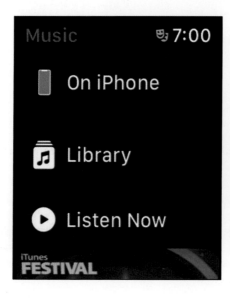

FIGURE 9-9: Listen to music on your Apple Watch without an iPhone.

Removing Music from Your Apple Watch

If you've grown tired of some songs on your Apple Watch, here's how to remove them:

1. **On your iPhone, open the Apple Watch app, and then tap My Watch ⇨ Music.**

2. **Tap Edit in the top-right corner of the screen.**

    You see only Edit if you have music synced to your watch.

3. **In the Playlists & Albums section, tap the red – icon next to any music that you want to remove and then tap Delete (see Figure 9-10).**

    You can also turn off any automatically added playlists that you don't want on your watch.

FIGURE 9-10:
It's just as easy to take songs off your Apple Watch as it is to add them, but both procedures require your iPhone.

Streaming Radio to Your Apple Watch

Beginning with watchOS 8, which debuted in late 2021, you can play radio stations in the Music app on your Apple Watch. No, the watch doesn't have a little radio receiver inside, but you can stream online radio stations to your wrist-mounted device. In fact, you have two ways to do that.

Radio is part of the Music app on Apple Watch, with three stations to choose from (at this writing): Apple Music 1, Apple Music Hits, and Apple Music Country. There is no cost to listen to these stations (see Figure 9-11). But additional stations are available through an Apple Music subscription.

FIGURE 9-11:
A look at the Radio tab inside the Music app on Apple Watch.

To listen to Apple Music Radio, follow these steps:

1. **Make sure that your Apple Watch is near your iPhone or connected to a Wi-Fi network (or connected to a cellular network, if you own an Apple Watch with cellular).**

2. **Open the Music app on your Apple Watch.**

   The icon has a red background with a white musical note.

3. **Tap Radio.**

4. **Tap a station: Apple Music 1, Apple Music Hits, or Apple Music Country.**

To listen to a featured or genre station, follow these steps:

1. **Open the Music app on Apple Watch.**

2. **Tap Radio.**

3. **Turn the Digital Crown button to scroll through stations and genres created by music experts.**

4. **Tap a genre to see its stations.**

5. **Tap a station to play it.**

Did you know that you can also listen to thousands of broadcast radio stations on your Apple Watch? Yes, and it's super-simple to do.

To listen to broadcast radio, follow these steps:

1. **Make sure that your Apple Watch is near your iPhone or connected to a Wi-Fi network (or connected to a cellular network, if you own an Apple Watch with cellular).**

2. **Ask Siri to play something.**

   You can ask for stations by name, call sign, frequency, and nickname. You might say "Hey Siri, play Wild 94.9" or "Hey Siri, tune in to ESPN Radio."

TIP You don't need a subscription to Apple Music to listen to broadcast radio.

Sharing content from the Music app

You also have the ability to share music with others right from the Apple Watch app. If you're loving a track and must share it with friends and family, then you can do it all with a couple of taps on your wrist. The following steps and Figure 9-12 show you how.

Here's how:

1. **When listening to a track in Library, on iPhone, or via the Radio feature, tap the dots in the lower right of the screen.**

   Now you see a few options tied to that song, such as Remove (not for Radio), Add to Playlist, Love (if you enjoy it), Suggest Less (if you don't), View Album, and more.

2. **Tap Share Song. If you're on Radio, you can also select Share Station.**

   Doing so brings up the last couple of names you've messaged with. Or scroll down to select Messages or Mail.

3. **Choose someone to send the song to and they'll receive a note with the name of the song and artist.**

Sharing is caring, amirite?

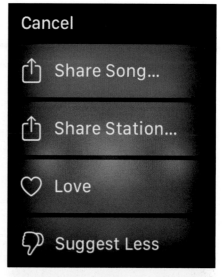

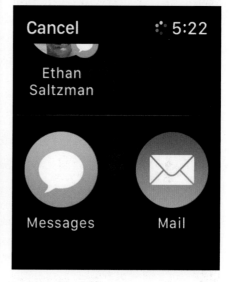

FIGURE 9-12:
Share the
songs you're
listening to
right from
within the
Music app. It's
super simple
to do so.

Playing Podcasts, Audiobooks, and Radio Plays

You can use your Apple Watch to enjoy many other kinds of audio entertainment, including podcasts, audiobooks, and radio plays. This section shows you how to pull it off.

Podcasts

Podcasts (a word that fuses *iPod* with *broadcasting*) are free downloadable shows from the Internet. There is a wealth of content to choose from, regardless of your tastes. Many podcasts have video, but in this section, I'm referring only to audio-based ones.

Unlike your favorite local radio station, podcasts let you to choose when the program starts (a process called *time shifting*) and where you want to listen to it, even in another country (a process called *place shifting*). Anyone can publish a podcast, from a 16-year-old video game fan or a huge media corporation such as ABC/Disney, CNN, or HBO. Thousands of radio stations have podcasts of their popular programs. Best of all, they're free.

Some podcast hosts are making a killing — and not just from advertisements or endorsements. Joe Rogan, for example, is reportedly raking in a whopping $200 million from Spotify to be exclusive!

Downloading podcasts

You can download podcasts to your computer (via Apple Podcasts on a Mac or iTunes on a Windows PC), an iPhone or iPad, and even your Apple Watch.

To download a podcast to your iPhone, follow these steps:

1. **Tap the Podcasts icon.**

    You'll see Listen Now, Library, and Search.

    Yep, new in 2022 is the ability to tap Search and discover and follow new podcasts right on your wrist — instead of using the iPhone as a middleman, so to speak (see below).

2. **Tap Library and you'll see podcasts divided into Saved, Downloaded, and Latest Episode.**

3. **If you like a podcast, you can tap to subscribe when prompted.**

    When you subscribe to a podcast, every time a new episode is available, it downloads automatically to your phone.

Now it's time to sync podcasts to your Apple Watch!

Syncing podcasts with your Apple Watch

This stuff is really easy, folks. In fact, all your Apple Podcasts subscriptions sync with your Apple Watch when it's charging! Syncing a few podcasts may take a few minutes, so be patient.

Playing podcasts

When you want to hear a podcast, tap the Podcasts app's icon (purple with a white symbol) on your Apple Watch Home screen. You see all your subscribed podcasts on the main Podcasts screen. Twist the Digital Crown button to scroll through the list if you subscribe to more podcasts than will fit on the screen. Figure 9-13 shows what a podcast looks like on Apple Watch.

FIGURE 9-13: A look at the Podcast app on Apple Watch, which especially looks nice on the larger Apple Watch Series 7 screen (center), while album artwork is on the left and an Audiobook on the right.

Alternatively, raise your wrist, and ask Siri to open a particular podcast.

Inside the Podcasts app, you can do the following:

>> **Move among your podcasts.** Flick the screen to navigate among your podcasts. If you swipe down, you see an option to play podcasts stored on your iPhone or the Library tab, which opens individual podcasts on your Apple Watch (see Figure 9-14).

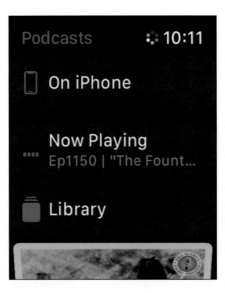

FIGURE 9-14:
When you're inside the Podcast app on Apple Watch (left), swipe up to access the library stored on your Apple Watch or on a nearby iPhone (right). Recognize this guy?

» **Play a podcast through a compatible speaker or TV.** Press and hold the screen (Force Touch,) and choose AirPlay.

» **Sync your podcast manually.** Podcast episodes are removed from your Apple Watch automatically after you listen to them, and podcasts that you subscribe to refresh automatically whenever new episodes are available. If you want to choose podcasts to sync manually, however, follow these steps:

a. *Open the Apple Watch app on your iPhone.*

b. *Tap My Watch ⇨ Podcasts ⇨ Custom.*

c. *Choose shows with episodes to sync to your Apple Watch.*

Figure 9-15 shows what this process looks like.

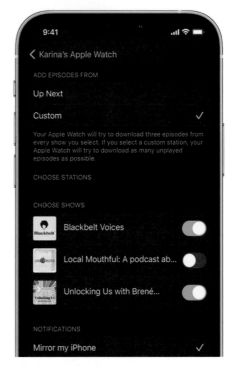

FIGURE 9-15:
If you don't want all podcasts to be synced with your Apple Watch automatically, open the Apple Watch app, and in Podcast Settings, choose what you want.

Audiobooks

You might call audiobooks today's answer to books on tape — if you're old enough to remember them! As you might expect, audiobooks are usually spoken versions of books read by a narrator as though you were being read a bedtime story. Audiobooks differ from radio plays, which I discuss in the next section, in that a cast acts out a performance, often with sound effects and music.

You can download millions of audiobooks from the Internet, but unlike podcasts and radio plays, they're usually not free. Still, they're a great way to have a book read to you if you're visually impaired, if you want to make a long commute in the car (or on the bus or train) more enjoyable, or if you have trouble reading and would rather hear someone read to you.

Audiobooks are available for new books (including *New York Times* bestsellers) as well as for slightly older and even classic titles. It doesn't matter whether the genre is fiction or nonfiction; chances are that you can find audiobook versions of many paperbacks and hardcover books, as well as many electronic books (e-books).

Some audiobook services (such as Audible) support Apple Watch, so you can download and listen to books on your wrist-mounted companion. Figure 9-16 shows the Apple Watch Audible option; you can find more information at Audible.com.

FIGURE 9-16: The easiest way to play an audiobook on Apple Watch is to use an app such as Apple's own Books (on iPhone) or a service such as Audible.

You have other ways to play audiobooks on your wrist:

>> **Copy a downloaded audiobook to a music folder in iTunes.** This process makes the file easier to control and play on an Apple Watch, but it might be a bit more difficult if the audiobook has digital rights management (DRM) encryption. DRM might limit or prevent the file from copying to another device. (DRM protects the copyright owner to help reduce piracy, which is the unauthorized distribution and/or duplication of copyrighted material.)

» **Download a free audiobook to Apple Watch.** Thousands of human-read or computer-generated audiobooks based on classic works are freely available in the public domain, and you can copy them to Apple Watch. The process takes some work, but you can copy these freely (and legitimately) available audiobooks to your Apple Watch from such websites as Archive.org (`https://archive.org/details/audio_bookspoetry`), Loyal Books (`www.loyalbooks.com`), and Project Gutenberg (`www.gutenberg.org/`).

Playing audiobooks

Here's the easiest way to play an audiobook on your Apple Watch:

1. **Launch the Books app on your iPhone, and tap the Explore Audiobooks tab.**

2. **Scroll to New & Trending, and peruse the Top Charts, search by keyword, and more.**

3. **Download an audiobook you want to listen to.**

4. **Tap the Audiobooks icon on Apple Watch.**

   You see all available audiobooks (see Figure 9-17).

5. **Choose how to listen.**

   The app offers On iPhone and Library options. Choose Library if you'd like to play Apple Books audiobooks directly from your Apple Watch.

Beginning with the watchOS 6 operating system, you no longer need to start an audiobook on your iPhone to continue listening on Apple Watch.

Syncing iTunes audiobooks with Apple Watch

After you've downloaded the audio files (usually MP3s) to iTunes on your Windows PC, follow these steps to sync them with your Apple Watch:

1. **Launch iTunes, and click the Music tab.**

   In the top-left corner, you should see a little icon with an arrow.

2. **Click this icon to display some options, including New.**

3. **Click New ⇨ New Playlist.**

4. **Name your audiobooks playlist.**

5. **Wherever you downloaded the DRM-free audiobooks to your computer — maybe your desktop or a Downloads directory — drag and drop them into this new playlist.**

   You're ready to sync your new playlist with Apple Watch so that you can hear your audiobook(s) without a nearby iPhone.

Radio plays

If you're listening only to music on your iPhone, iPad, or Apple Watch, you're missing out on thousands of free dramas and comedies that can keep you entertained on the go.

Popularized in the 1940s before TV took off, radio plays — or old-time radio (OTR) shows, as they're often referred to today — are enjoying a 21st-century revival thanks to the Internet and MP3s. A whole new generation of listeners is experiencing these well written, wonderfully performed "theater of the mind" episodes.

Recommended shows include the creepy *Inner Sanctum* and *The Price of Fear* (with Vincent Price) mysteries; nail-biting adventures from *Suspense* and *Escape*; the hilarious antics of Jack Benny and those of Abbott and Costello; and sci-fi classics such as *Journey Into Space* and *X Minus One* (featuring many Ray Bradbury yarns).

Filling your digital devices with these timeless tales is as easy as subscribing to one of the many dozen OTR podcasts (some with daily updates) or by bookmarking such websites as Archive.org (`https://archive.org`), Relic Radio (`https://www.relicradio.com/otr`), and OTRCat.com (`https://www.otrcat.com`), each of which has thousands of free downloadable episodes.

There are also excellent podcasts devoted to OTR, such as the Mysterious Old Radio Listening Society (`ghoulishdelights.com/series/themorls`).

Because most of these older shows have copyrights that have long since expired (or didn't have any to begin with), they're available for free through the public domain. Some newer programs, such as many BBC radio plays and recently published radio dramas based on *The Twilight Zone,* are still protected by copyrights that prohibit you from copying and distributing them. When in doubt, contact the website that houses these audio plays.

To sync radio plays with your Apple Watch, follow these steps:

1. **Launch the Music app and then tap the Playlist tab.**

2. **Tap the New Playlist button (+).**

3. **Give a name to your radio-shows playlist.**

   You might call it Radio Shows, OTR, Inner Sanctum, or Suspense, for example.

4. **Tap Add Music ⇨ Library.**

5. **Select the folder that contains the radio plays.**

   You're ready to sync your new playlist to Apple Watch so that you can hear your radio plays without a nearby iPhone.

Now you can look forward to — rather than dread — your daily commute.

Controlling Apple TV and Apple Music (or iTunes)

Your fancy-schmancy new Apple Watch can control your iPhone wirelessly. It can play music or snap the shutter on your iPhone's camera (see Chapter 12). More important, it can manage the Apple TV box connected to your TV set or the iTunes software installed on a Windows PC. (On a Mac, iTunes is now called Apple Music.)

For iTunes, the Remote app lets you play back content on your computer as though it were a TV as long as you have media in your iTunes library. So put that mouse away, and start controlling iTunes from your wrist! If you're cooking in the other room and want to change the music pumping from the speaker, you don't need to go to your laptop or desktop.

Before you can control iTunes or Apple TV, though, you need to set up Home Sharing in iTunes and sign in with your Apple ID.

Setting up Home Sharing

To set up Home Sharing in iTunes on a Windows PC, follow these steps:

1. **Download the free Remote app from the App Store.**

2. **On your computer, open iTunes, and click the little rectangular icon in the top-left corner of the screen.**

3. **Click Preferences.**

4. **Click the blue Sharing tab, and select the media you'd like to share via Home Sharing.**

   Options include Music, Movies, Home Videos, TV Shows, Podcasts, iTunes U, Books, and Purchased.

5. **Connect your devices (such as an iPad and iPhone) to your home Wi-Fi network.**

6. **Sign in to Home Sharing via the Remote app.**

   Home Sharing is enabled in iTunes.

Controlling Apple TV remotely

To control Apple TV from your Apple Watch, follow these steps:

1. **Press the Digital Crown button to go to the Home screen.**

2. **Tap the Remote app.**

   You can also raise your wrist and say "Hey Siri, Remote." Either action launches the Remote app. You're presented two options: Apple TV and iTunes.

3. **Select Apple TV if you own one of these media boxes and it's attached to your TV.**

   As long as your iPhone and Apple TV box are joined to the same Wi-Fi network at home or work, you can access your Apple TV as though your Apple Watch were the small remote control that comes with it. (You can also control Apple TV with the Remote app for iOS — for the iPhone, iPod touch, and iPad — if you'd like to.)

4. **Tap one of the four arrows on the watch — Up, Down, Right, or Left — to navigate Apple TV's menus.**

   You don't have to reach for the hardware remote, phone, or tablet anymore; now you can control your media with a tap on your wrist.

5. **Tap the center Play/Pause button to stop and start your media at your convenience.**

   You're at home, with your feet up on the coffee table, and you're enjoying a bit of Netflix on your big screen — until your dog brings you a leash in its mouth (and with those sad eyes). Tap the center of your Apple Watch to pause playback and then take your pup for a walk.

6. **Press the Digital Crown button to return to the Home screen.**

REMEMBER

Be sure to turn off your TV and Apple TV box if you're done. Don't waste power and money.

Controlling iTunes remotely

To control iTunes (on a Windows PC) from your Apple Watch, follow these steps:

1. **Tap the Digital Crown button to go to the Home screen.**

2. **Tap the Remote app.**

   You can also raise your wrist and say "Hey, Siri, Remote." Either action launches the Remote app.

3. **Inside the app, tap iTunes.**

   You're prompted to enter a four-digit code to access your iTunes library.

REMEMBER

   You need to be connected to the same Wi-Fi network as your Windows PC and be signed into Home Sharing in iTunes (covered earlier in this section).

4. **Swipe around the Apple Watch screen to access the library of content on iTunes, and select something to play.**

   The Remote app lets you control your iTunes library from anywhere in your home. Fast-forward, pause, or skip back a track or two. Choose a song, shuffle an album, or select a custom playlist. And if your Mac or PC is sleeping, opening the Remote app wakes it up.

IN THIS CHAPTER

» Using Apple Pay on Apple Watch

» Setting up Apple Watch for Apple Pay

» Paying with Apple Watch without a nearby iPhone

» Accessing Wallet on Apple Watch (including passes)

» Adding car and home keys to Wallet

» Adding COVID-19 vaccination proof in Apple Wallet

» Using Apple Cash and Apple Cash Family on Apple Watch

» Using Apple Watch for deals and rewards

» Mastering the Home app to monitor and control your home

Chapter **10**

Making Mobile Payments with Apple Watch Controlling Your Smart Home

This chapter looks at how to use Apple Pay on Apple Watch, as well as Apple Pay Cash, to send friends money. First, though, a brief primer on what Apple Pay is.

If you own an iPhone 6 or newer, then perhaps you've had a chance to try out Apple Pay (www.apple.com/apple-pay), Apple's proprietary mobile payment solution that lets you buy goods and services without needing your wallet.

Simply wave your compatible phone over a contactless terminal at a participating retailer — with your finger on the Touch ID sensor (built into the Home button) or with newer models, looking at your iPhone's camera using Face ID technology — and the transaction is completed. This saves time at retail because you don't have to dig for the exact change or deal with finding the right credit or debit card in your wallet.

Intrigued? Read on to find out how to set up and use Apple Pay, how to use Apple Pay Cash, how to send cash to friends, and how to access Apple Watch-specific deals and rewards.

Apple Watch and Security

Before I talk about paying for something on Apple Watch, a word on Apple Pay on iPhone.

If you're worried about security, Apple Pay (shown in Figure 10-1) doesn't disclose your financial information to the retailer; it uses a unique numerical *token* rather than actual credit or debit card numbers shared with the store (which could put your info at risk). In fact, Apple claims using Apple Pay is safer than using a physical card.

Because of the integrated near field communication (NFC) antenna in the latest iPhones, you don't need to open an app or even wake up your phone to use it. As long as your one-of-a-kind fingerprint or face is detected — two examples of biometrics technology that prove it's you and only you — you should feel a subtle vibration on the iPhone, and hear a small beep to confirm the *digital handshake* has been made.

You can also use Apple Pay to pay for apps in the online App Store. For this, you can also use the Touch ID or Face ID sensor built into iPhones, iPads, and MacBook Pro (but you can't use the tablet or computer to pay at retail). Checking out is as easy as selecting Apple Pay from the list of options and using your finger or face to identify you.

Before you can use Apple Pay — on any device — you first need to establish your Apple Pay account. You do this first on iPhone and then Apple Watch.

FIGURE 10-1:
Leave your wallet at home and use the Wallet app instead! Simply tap your iPhone or Apple Watch at retail to buy something securely.

Setting Up Apple Pay on iPhone

Setting up your iPhone for Apple Pay involves your Wallet app (formerly Passbook), which stores such things as boarding passes, movie tickets, store coupons, loyalty cards, COVID-19 vaccination proof (in most states), and more. It can store your credit and debit card information too.

I get to setting up Apple Pay on Apple Watch shortly, after covering the other Apple devices first. To set up Apple Pay on your iPhone, follow these steps:

1. **Tap the Wallet app icon on your iPhone.**

   The Wallet app launches. The first time you open it, you're prompted to allow Wallet to use your location. Select what you're comfortable with. I would recommend selecting "Allow While Using App."

2. **Tap the + sign in the upper-right corner, as shown in Figure 10-2.**

   This lets you add a new card.

3. **Use your device's main (rear-facing) camera to capture the information on your credit and debit card or choose to enter it in manually.**

   You have to fill in additional information it might ask for, such as the security code.

4. **Your bank then verifies your information and decides if you can add your card to Apple Pay.**

Not all banks are supported, but many are. You may be prompted to provide additional information to verify it's you for your own security and privacy.

5. **Tap Next after your card is verified.**

You can now start using Apple Pay at supported retailers and in the App Store. Keep in mind that the first card you add is your default payment card, but you can go to the Wallet app anytime to pay with a different card (or select a new default in Settings ⇨ Wallet & Apple Pay). Otherwise, you don't need to do anything. See the "Looking at the Wallet App" section later in this chapter for more on changing your default card and other options.

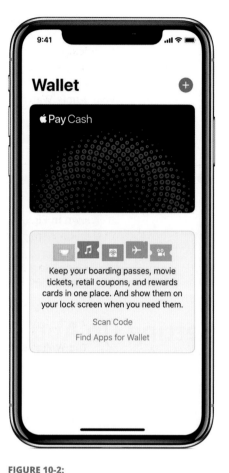

TECHNICAL STUFF

Whenever you use your credit or debit card at a retailer, your card number and identity, including your name, is visible to the retailer. This can cause privacy and security risks. However, Apple Pay uses a unique Device Account Number (DAN) for each transaction. This DAN is generated, encrypted (securely blocked from outsiders), and then stored in the Secure Element — a dedicated chip inside the iPhone, iPad, and Apple Watch. Apple says these numbers are never stored on Apple servers. The DAN and a dynamically generated security code are used to process your payment when purchasing something; therefore, your actual credit or debit card number is never shared by Apple with the retailer.

FIGURE 10-2:
It's a breeze to set up your iPhone with Apple Pay. The step-by-step instructions should take you less than three minutes (seriously).

What happens if you lose your iPhone, iPad, or Apple Watch? Don't worry about them being used to make a fraudulent purchase. The crook will need your unique finger on the Touch ID sensor (Home button), your face for Face ID, or a PIN for

Apple Watch. If you ever misplace your phone or tablet, you can use the Find My iPhone feature on the iCloud website (www.icloud.com) or on another iOS device to quickly put your device in Lost Mode to suspend Apple Pay. Alternatively, you can completely wipe your device clean of any data. And yes, you can now find your Apple Watch using the Find My app on iOS devices (and iCloud.com). So long as your Apple Watch has power, it will show up on the Find My map on an iPhone or iPad.

Setting Up Apple Pay on Apple Watch

As shown in Figure 10-3, the process to set up Apple Pay on Apple Watch is similar to the other Apple products:

1. **Open the Apple Watch app on your iPhone, and tap the My Watch tab (lower-left corner).**

   If you have multiple watches, choose one (for me it says "Marc's Apple Watch, 44mm").

2. **Tap Wallet & Apple Pay near the bottom of the screen.**

3. **Tap the orange words Add Card, and follow the steps to add a card.**

4. **To add a new card, tap Add Credit or Debit Card.**

   If you're asked to add the card that you use with iTunes, cards on other devices, or cards that you've recently removed, choose them, and then enter the card security codes. ***Note:*** Apple Pay requires a passcode on Apple Watch. If you don't have one set up, you can tap the orange "Set Up Passcode" and follow the prompts.

5. **Tap Next.**

   Your bank or card issuer will verify your information and decide if you can use your card with Apple Pay. If your bank or issuer needs more information to verify your card, you're prompted for it. When you have the info, go back to Wallet & Apple Pay and tap your card.

6. **After your bank or issuer verifies your card, tap Next.**

   You can now start using Apple Pay (see the next section).

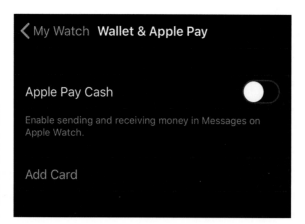

FIGURE 10-3:
You only need
to activate a
credit or debit
card once on
your Apple
Watch, then
you're good
to go.

Using Apple Pay with Your Apple Watch

Because you're wearing Apple Watch as opposed to holding it, Apple Pay on Apple Watch is even easier and faster to use at retail than an iPhone. Simply wave your Apple Watch over a contactless terminal at a supporting retailer.

Those wearing an Apple Watch can keep their phones tucked away in a purse or pocket — or even left at home if preferred — but the payment process is similar.

To pay for something with your Apple Watch, follow these steps:

1. **Step up to a contactless reader (point-of-sale terminal).**

   Yes, these are the same ones that support smartphones and NFC-enabled cards, such as Visa's PayWave, MasterCard's PayPass, and Amex's ExpressPay. The person behind you might not be sure what you're doing. Just smile and proceed to the next step.

2. **Double-tap the side button on Apple Watch.**

   You're now ready to make the *digital handshake*.

3. **Hold the face up to the contactless reader, and within a second or two, a tone and slight vibration confirms your payment information has been successfully sent.**

   You won't need to open an Apple Pay app on the watch or anything like that. No wonder Apple calls it "Your wallet. Without a wallet."

4. **If you need to choose Credit or Debit on the terminal, choose Credit.**

   Depending on the store and transaction amount, you might need to sign a receipt or enter your PIN.

You will, though, have to open the Wallet app on Apple Watch if you want to switch cards while paying. To do that, follow these steps:

1. **Double-tap the side button to display Wallet.**

   You can also say "Hey Siri, Wallet" or press the Digital Crown button and find and tap the Wallet app. This is a similar process to paying using Apple Pay, but don't hold your watch up to the contactless reader just yet.

2. **Swipe up or down with your fingertip or twist the Digital Crown button to browse through your cards.**

3. **Select the card you want.**

4. **Hold the Apple Watch face near the NFC reader to pay. See Figure 10-4.**

   You should feel a slight vibration and hear a faint chime to confirm the transaction was successfully completed.

FIGURE 10-4: At a contactless reader, tap your iPhone or Apple Watch to use Apple Pay (left). On Apple Pay for Apple Watch, press the side button twice and tap to pay (right).

REMEMBER

Apple Watch works with any iPhone 5 or newer; iPad Pro, iPad Air, iPad, and iPad mini models with Touch ID or Face ID; and Mac models with Touch ID or any models introduced in 2012 or later with an Apple Pay-enabled iPhone or Apple Watch nearby. You can use Apple Pay with a compatible Mac for payments on the web (with the Safari web browser).

Paying without a Nearby iPhone

Remember, Apple Pay differs from most other mobile payment solutions because your credit or debit card number is never visible to the retailer and is thus safer to use. When you add your card information to Apple Watch, that unique and encrypted Device Account Number (DAN) is assigned and stored on a dedicated chip inside the watch (that Secure Element technology, like with the newer iPhones and iPads).

"But wait a sec," you're thinking. "Apple Pay on iPhone requires my fingerprint on the Touch ID sensor or my face for Face ID (for iPhones that don't have a Home button), so how do I do that on Apple Watch?"

To enable Apple Pay on the watch, you need to create a four-digit passcode using the companion Apple Watch app on your iPhone. This passcode is used to author-ize Apple Pay whenever you put the watch on your wrist. Go to My Watch ⇨ Passcode to enable and create a passcode using the virtual keyboard, as shown in Figure 10-5.

And as you might guess, those sensors on the back of the watch aren't just used for your heartbeat; they *know* whenever the watch has been taken off. And you must once again type your secret passcode when you put the Apple Watch back on your wrist.

Clever, eh?

That way, if someone puts on your Apple Watch or tries to use it without slapping it on his or her wrist, your information is safe because the other person won't know your passcode.

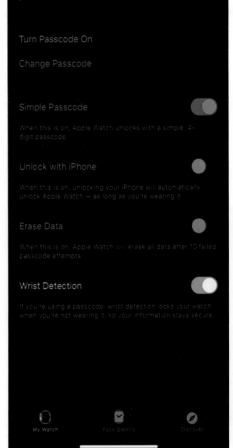

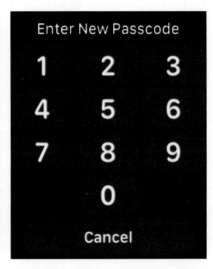

FIGURE 10-5:
Use your
iPhone to
create your
four-digit
passcode for
your Apple
Watch (left).
You are
prompted to
enter one
on a virtual
keyboard
(right).

Looking at the Wallet App

Just like the app on your iOS devices — iPhone, iPad, or iPod touch — Wallet is available on Apple Watch, and it also lets you store your airline boarding passes (with a scannable QR code), movie and sports tickets, coupons, loyalty cards, vaccination proof, and much more, as shown in Figure 10-6. Think of it as a handy digital version of your actual wallet (if you still use one).

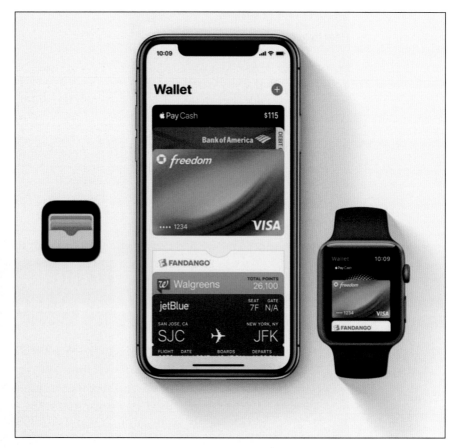

FIGURE 10-6:
The Wallet
app on Apple
Watch and
iPhone for
retail
purchases
can also store
plane tickets,
loyalty cards,
event passes,
and even
coffee
(Starbucks)!

Other sections in this chapter talk about how Wallet can house your payment cards — credit or debit — for use with Apple Pay and how to change between cards if you like. In this section, I cover other things you can do with the Wallet app.

After all, accessing Wallet from your wrist is awesome. For example, you can use it to have an airline attendant scan a QR code on your wrist to board a plane while you're negotiating an important deal on your iPhone. Plus, location and time services allow Wallet to notify you about relevant information just when you need it, such as gate changes at the airport.

When you download an app that supports Wallet, you have the option to add it to Wallet. Then, in your Wallet, you can flip through these apps, represented by virtual cards, with your fingertip to see each of their services and get them ready for when you need them. You can also be notified when you step inside a supported retailer or another location, or when information regarding a particular service updates, such as a cancelled concert.

How to get going?

Wallet is already preinstalled on all iPhones. If the app you download supports it, Wallet asks you if it can add that particular app to your Wallet. Examples of sup-ported apps include United Airlines, Target, Ticketmaster, Fandango, Walgreens, and Starbucks, to name a few.

For example, to use the Starbucks app, follow these steps on your iPhone:

1. **Download the Starbucks app from the App Store.**

2. **Sign up for the Starbucks Rewards loyalty program or sign in if you're already a member.**

   You're prompted with this message: "Would you like to add your Starbucks card to your Wallet?"

3. **Tap Continue to add Starbucks to Wallet.**

   This message appears: "Choose your favorite stores, and your Starbucks card will appear on your lock screen when you arrive."

4. **Let the phone's GPS locate the nearest Starbucks, and then you can select a few as favorites.**

   When you're done, you should see the Starbucks card pop up in your Wallet app — with a scannable QR code and cash balance of your account, as shown in Figure 10-7. Now hold out your Apple Watch to have the screen scanned by a barista at the cash register.

FIGURE 10-7: Starbucks is one of the more popular Wallet-supported loyalty cards. This is what the Starbucks app looks like on your Apple Watch.

Adding passes to Wallet

You can add passes in many ways — as well as student ID cards and virtual car keys — to your Wallet app. Let us count the ways, beginning with adding it to your iPhone (which then syncs with Apple Watch):

>> **Find through the Wallet app.** Open the Wallet app, tap Edit Passes, and tap Find Apps for Wallet.

>> **Scan a barcode or QR code.** Open the Wallet, tap Edit Passes, tap Scan Code, and scan with your iPhone camera.

>> **Add through a notification.** If you use with Apple Pay at a supported merchant and you get a notification, tap the Wallet notification to add the pass to Wallet.

For student ID cards, use your school's student account management app.

For car keys, use the app provided by your car's manufacturer, add keys from email, or use your car's information display (if your vehicle is supported).

Note: In some cases, you might need to tap Add to Apple Wallet, and then tap Add in the upper-right corner of the pass. Or you might see a pop-up notification with an Add button you can tap to add your pass to Wallet.

To add passes to Wallet in Apple Watch, you can

>> Follow the instructions in the email sent by the pass issuer.

>> Open the app from the issuer.

>> Tap Add in a notification from the issuer.

To remove passes from Wallet, on your iPhone, open the Wallet app, and scroll down to tap Edit Passes. Tap the arrow button. Tap Delete. When you tap Delete, the pass is also removed from the Wallet on your Apple Watch.

Using passes on Apple Watch, including the Car Key feature

Now, whenever you'd like to call up your pass in Wallet, double-press the side button on Apple Watch and select what pass you'd like to use from the available cards. You can scroll up or down and select one, such as a Starbucks code, Student ID card, transit pass, and so on.

Let me share with you one of the coolest ways to use your Apple Watch: as a virtual car key!

That's right, car keys have been around for about 100 years. But did you know you can now leave them at home and just use your Apple Watch or iPhone?

If your vehicle offers the option to control your vehicle with a digital car key, you can add this key to the Wallet app.

Then, when you want to get into your vehicle, place your Apple Watch or iPhone near the handle to unlock it, via NFC (near-field communication). Climb inside, place your iPhone in the car's key reader, or hold your Apple Watch near the reader, and then press the car's Start button to drive. See Figure 10-8.

FIGURE 10-8:
While it may sound like science-fiction, you can use your Apple Watch to unlock and drive your (supported) vehicle, such as this BMW. You can also share a virtual key with family and friends, too.

If you own an iPhone with Ultra Wideband (UWB) technology (iPhone 11 and newer, excluding Apple Watch SE), or you're wearing an Apple Watch Series 7 (2021) model, they both have Ultra Wideband (UWB) technology, so you don't need to hold the device up to the vehicle handle.

TIP

How to add car keys to Apple Watch

Before you begin, make sure that your car supports a digital key feature. You may need to contact the manufacturer.

1. **Open the car manufacturer's app and follow the instructions to set up a key.**

2. **Add the car keys to the Wallet app on your iPhone that's paired with your Apple Watch.**

3. **Open the Watch app on your iPhone.**

4. **In the My Watch tab, scroll down and tap Wallet & Apple Pay.**

5. **Tap Add next to the card for your car key.**

Cool, no? It gets better.

One of the best features of a digital key is the ability to share it. Through iMessage you can select a family member to grant them access — with restrictions if you want.

How to share a digital car key

Want to send someone a virtual key? It's done on your iPhone, and car key sharing invitations can only be sent via iMessage.

1. **Open the Wallet app on your iPhone.**

2. **Tap the card for your car key.**

3. **Tap the More button.**

4. **Tap Invite.**

5. **Tap Set Access, then select the level of restrictions that you want to place on the shared car key.**

6. **Tap Invite. A new message appears.**

7. **Begin typing a person's name in the message's To field, then tap their name when it appears. If the recipient isn't in your contacts, you can type a phone number.**

8. **Tap the Send button.**

And yes, if you misplace your Apple Watch, no one can access the Wallet app unless it's you (because of your passcode), but you can also turn off the Car Key remotely through iCloud.

Again, if you own Apple Watch Series 7, your digital car key recognizes you as you approach your vehicle. Unlock the door, activate the alarm, or access climate controls — all from a distance — thanks to the Ultra Wideband technology in this particular model of Apple Watch.

Storing a digital Home key to Wallet

Apple Watch now lets you store a digital key to your home, providing a convenient (yet still secure way) to get into your home, as shown in Figure 10-9. This feature is particularly handy for sharing a digital key with family, guests, or even a contractor.

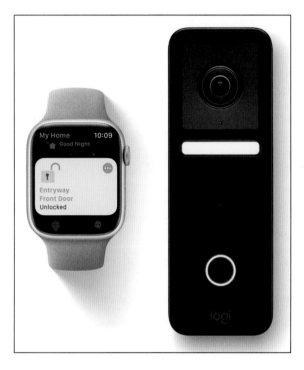

FIGURE 10-9:
You can now assign a Home key to Apple Watch, which works with a number of HomeKit-enabled door locks.

Home key works with a variety of popular lock brands, and allows you to open multiple locks with a single key.

You will need a compatible HomeKit door lock, and you'll get inside your door by placing your iPhone or Apple Watch near the lock.

To set up a home key, follow these steps:

1. **Add the lock to the Home app on your iPhone.**

2. **Choose an unlocking option:**

   - **Enable Express Mode:** Unlock the door just by holding your iPhone near the lock. To enable Express Mode on your Apple Watch, after adding the home key, open the Settings app on your Apple Watch, go to Wallet &

Apple Pay ⇨ Express Mode, and then turn on Express Mode for your home key.

- **Require Authentication:** Similar to paying for a purchase with your iPhone or Apple Watch, double-click the side button and then hold your device near the lock.

Choose automations such as Lock After Door Closes and Lock When Leaving Home.

Adding COVID-19 vaccination proof in Apple Wallet

Now that we're returning to "normal" after the global pandemic of 2020-22, some businesses want to see proof of vaccination before letting people in their doors. In fact, it may be the law, depending on where you live. As such, people who want to go to restaurants, gyms, and sports arenas want to keep their vaccine passport on their iPhone and Apple Watch.

Beginning with iOS 15.1 for iPhone, Apple lets you do just that, making your vaccination proof a "card" and even a QR code.

If you received a QR code (on paper or scannable on a computer) follow these steps and see Figure 10-10:

1. **Open the Camera app from the iPhone's Home Screen.**

   Make sure your rear-facing camera is activated (not your self-camera).

2. **Hold your device so that the QR code appears in the viewfinder in the Camera app.**

   Your device recognizes the QR code and shows a Health app notification.

3. **Tap the Health app notification.**

   If it's a verifiable vaccination records, you will be prompted for the next step.

4. **Tap Add to Wallet & Health to add the record to the Health app and Wallet app.**

   Now tap Done.

5. **Open Apple Wallet on your Apple Watch and you'll be able to access the card here, too, which can be scanned by someone.**

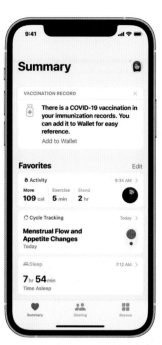

FIGURE 10-10:
Scan your
government-
or healthcare
provider-
mandated
vaccination
proof on a
piece of paper
or website
using your
iPhone camera
to add it to the
Health app and
Apple Wallet,
which can then
be accessed on
your watch.

If you received a downloadable file from your health provider or government:

1. **Tap the download link on your iPhone or iPod touch.**

2. **For verifiable vaccination records, tap Add to Wallet & Health to add the record to the Health app and Wallet app.**

3. **Tap Done.**

4. **Open Apple Wallet on your Apple Watch to access your vaccination passport.**

Seeing notifications for passes on Apple Watch

To get notifications from Wallet, including when a pass changes — for example, the gate on your boarding pass is now at A47 instead of A49 — follow these steps:

1. **Open the Apple Watch app on your iPhone.**

2. **Select the My Watch tab, and then scroll down and tap Wallet & Apple Pay.**

3. **Tap Mirror My iPhone below notifications.**

Using Apple Pay Cash on Apple Watch

So far, this chapter has discussed Apple Pay and how to use your Apple Watch or iPhone to make payments at retail locations or online. Did you know you can also send people money through your Apple device?

Called Apple Pay Cash, this easy-to-use technology is designed for person-to-person payments. It's essentially a version of Apple Pay that lets you send or receive money right in the Messages app (discussed in Chapter 5), or by asking Siri.

You can use the cards you already have in Wallet to send money, split a bill at a restaurant, or chip in on a gift for a coworker. Figure 10-11 shows this wonder in action.

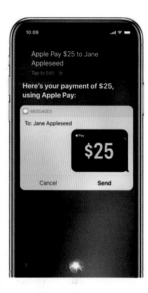

Apple Pay Cash requirements

Here's what you need to get going:

>> United States residency. You also need to be 18 years of age or older.

>> A compatible device with iOS 11.2 and later or watchOS 4.2 and later.

>> Two-factor authentication for your Apple ID, which requires both a password and one-time code sent to your iPhone to confirm it's really you.

>> Connection (sign in) to iCloud with your Apple ID on any device that you want to use to send or receive money.

>> An eligible credit or debit card in Wallet (see earlier in this chapter on adding cards).

>> Agreement with the Terms and Conditions, something required the first time that you try to send or receive money.

>> Verification of your identity (you may be asked this).

Now, right from your iPhone, iPad, or Apple Watch, friends can securely send and receive money. Just be aware that Apple Cash is for personal use only and not for any business–related activities, such as paying business expenses or paying employees, per Apple.

Setting up Apple Pay Cash in Wallet

If you're setting up Apple Pay Cash for the very first time, you need to use any supported device that allows you to sign in to iCloud with your Apple ID. To do so:

1. **Tap Settings ⇨ Wallet & Apple Pay.**

2. **Tap the Apple Pay Cash card, and then follow the onscreen instructions.**

3. **To enable Apple Pay Cash on Apple Watch, open the Apple Watch app on iPhone, and then tap My Watch ⇨ Wallet & Apple Pay (see Figure 10-12).**

4. **Tap the Apple Pay Cash toggle so it turns green and says "Enable sending and receiving money in Message on Apple Watch."**

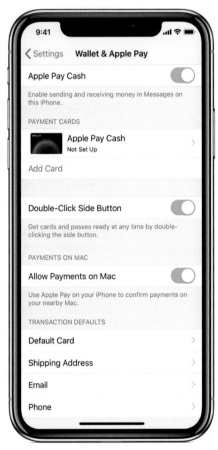

FIGURE 10-12: Setting up Apple Pay Cash on iPhone (left) and enabling Apple Watch support (right).

TIP

If you turn off Apple Pay Cash for any one device, such as on an iPad, you can still use Apple Pay Cash on other devices where you're signed in with your Apple ID, such as on an iPhone or Apple Watch.

Apple confirms there is no fee to use Apple Pay Cash with a debit card. If you send money using a credit card, though, there's a standard 3 percent credit card fee on the amount.

If someone sends you money via Apple Cash, it's automatically kept on your Apple Pay Cash card. You see your new Apple Pay Cash card in your Wallet, and you can use the money to send to someone; to make purchases using Apple Pay in stores, within apps, and on the web; or to transfer it from Apple Pay Cash to your bank account.

Using Apple Watch to send cash to friends

Once you're all set up, it's super easy to send cash to someone through your Apple Watch! Just follow these steps and see Figure 10-13:

1. **Open the Messages app (see Chapter 5 for more on messaging).**

   You may need to press Digital Crown to see all your apps.

2. **Start a new conversation or tap to open an existing one.**

3. **Scroll down, and then tap Apple Pay.**

4. **Choose an amount by turning the Digital Crown.**

   To enter an exact amount, tap the dollar amount, tap after the decimal, then turn the Digital Crown.

5. **Tap Pay.**

   To review the payment information or cancel, scroll down. Your money in Apple Pay Cash will be used to pay first.

6. **To send the payment, double-press the side button.**

7. **To cancel a payment after sending, tap the payment to see its details, and then check the Status field and tap Cancel Payment.**

TIP

You can use Siri to send money! Say something like, "Apple Pay 50 dollars to Steve for movie tickets" or "Send 25 dollars to Susan."

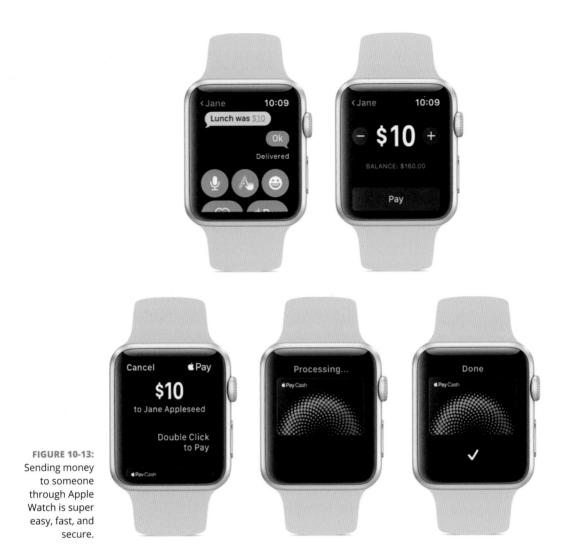

FIGURE 10-13:
Sending money
to someone
through Apple
Watch is super
easy, fast, and
secure.

Using Apple Watch for Other Deals and Rewards

Besides using Apple Watch to make purchases and offer support for loyalty cards, you can also use it in ways that serve up a richer shopping experience. Folding in other technologies, you can utilize Apple Watch to customize the store experience to suit individual preferences. For example, stores equipped with Apple's iBeacon location-based technology (`developer.apple.com/ibeacon`) can identify you from your Apple Watch (and/or nearby iPhone). This means the technology welcomes you, rewards you with points for stepping into a store, notifies you about deals, and gives personalized suggestions based on previous interests.

Of course, these are things you have to opt into. That is, you must first give the store permission in its downloadable app.

Imagine stepping into a retailer and seeing what you need to know on the Apple Watch screen. It's still early days, but the technology is there. Just ask Cyriac Roeding, the celebrated entrepreneur, investor, and cofounder/CEO of Shopkick, an app that helps you find deals and earn rewards.

"While it's still a little early, smartwatches like Apple Watch are a next, natural step in the evolution of the shopping experience," believes Roeding, whose Shopkick app is the most widely used shopping app at retail stores in the United States, according to Nielsen. As of April 2017, Shopkick said it had powered more than "200 million store visits, over 270 million product scans in aisle, and over $2.5 billion in total sales from brand and retail partners" (according to Wikipedia). Roeding says Shopkick's Apple Watch app offers much of the same experience as the smartphone app "but without you having to take out your phone." When you walk into one of the many stores with Shopkick's shopBeacon transmitters — built upon Apple's location-based iBeacon technology — you're welcomed by name, rewarded with redeemable *kicks* (points) for walking in the door, and notified of curated deals based on your previous shopping habits. See Figure 10-14 for an example of Shopkick's technology (courtesy of Shopkick).

FIGURE 10-14: A look at Shopkick's shopBeacon transmitter on the wall of a retailer.

shopBeacon's location technology, which uses ultrasound and Bluetooth low energy (BLE), does not collect any information about you, assures Roeding, because it only sends information instead of collecting data.

Mastering the Home App to Monitor and Control Your Smart Home

The Home app on Apple Watch provides a secure way to control "smart home" accessories, such as lights, door locks, smart TVs, thermostats, window shades, and smart plugs. The only catch is these devices must be compatible with Apple's HomeKit platform, which is gaining in popularity.

What's really cool is you can also see video streams on your wrist from a compatible doorbell camera, and send and receive intercom messages on supported devices. How convenient.

You'll first open the Home app on your iPhone, and the setup assistant will help you create a "home." Once you do that, you can define rooms, add HomeKit-enabled accessories, and create scenes, which allow you to control multiple accessories at once.

As an example of a scene, you might create a "Reading" scene that adjusts the lights, plays soft music on HomePod, closes the drapes, and adjusts the thermostat!

All the accessories, scenes, and rooms that you add on your iPhone are available on your Apple Watch.

TIP

Beginning with watchOS 8, you can automatically make suggestions for others nearby when one of your smart devices is activated. For example, if someone rings your video doorbell, you might see options like unlocking the door or turning on the entry lights, as shown in Figure 10-15. Over the years new Home features were added to your Apple Watch, including support for connected doorbells. In fact, you can access all your cameras in the new Camera room, with several aspect ratios supported (as in two-way audio on cameras that offer it). To see a fullscreen view, just tap.

Kids who wear an Apple Watch (in Family Setup mode) can now access compatible smart home items via Apple Watch, such as locking the front door and enabling a home alarm before going to sleep.

View your home status

The Home app on Apple Watch shows you the status of accessories you're currently using, such as the temperature of your thermostat or whether your front door is locked or unlocked.

FIGURE 10-15:
With
supported
video
doorbells,
see who's at
your front
door . . . right
on your wrist!

To view your status:

1. **Open the Home app on Apple Watch.**

2. **Tap any of the round buttons that appear just below your home's name.**

When multiple accessories appear in a status, tapping the status lets you control each device or set of grouped devices.

As you can see in Figure 10-16, the current status of all your devices is now conveniently displayed at the top of the Home app screen. Instantly see if your cameras or lights are on, whether batteries need recharging, if your software needs updating, and more.

Control smart home accessories and scenes

Apple lists "relevant" scenes and accessories near the top of the screen, such as a coffee maker in the morning or setting the alarm before you go to sleep at night.

To see the rest of your accessories, simply scroll up with your fingertip, and you can access Cameras, Favorites, or a specific room.

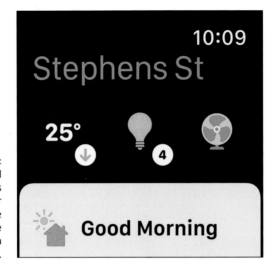

FIGURE 10-16:
The redesigned
Home app lets
you see your
compatible
smart home
devices at a
glance.

To control an accessory, do any of the following:

- » **Turn an accessory on or off.** Tap the name of the accessory, such as a bedside lamp. To unlock a compatible lock, tap the Home key.

- » **Adjust an accessory's settings.** Tap the More button (three dots) for an accessory. Tap Done to return to the list of accessories.

  Be aware the available controls depend on the type of accessory you're trying to access. A thermostat, for example, may let you adjust the temp and other settings, while smart light bulbs may offer controls for brightness and changing colors. To see additional controls, swipe left.

- » **Control favorite accessories or accessories in a room.** Tap Favorites or a room, then tap an accessory or tap the More button to adjust the accessory's settings.

- » **View a camera's video stream.** Tap Cameras, then tap a camera.

To run a scene, open the Home app on your Apple Watch, then tap the scene.

4

More Apple Watch Tips and Tricks

Customize your Apple Watch by adding third-party apps, such as ESPN, Twitter, TripAdvisor, NPR One, and Fandango. You can download apps directly to Apple Watch over the air or sync through your iPhone.

Explore how to sync photos and download apps to your Apple Watch as well as how to establish and modify settings in the companion Apple Watch app on your iPhone.

Use Apple Watch to unlock your Mac, or to view verification codes on Apple Watch.

Discover how to turn your Apple Watch into a camera, including storing and viewing photos from your iPhone, as well as using the watch as a remote control for the iPhone's camera.

Enhance your Apple Watch experience by downloading gaming apps, including such iPhone favorites as Trivia Crack and Best Fiends.

IN THIS CHAPTER

» Customizing the Apple Watch
   experience with apps

» Downloading and installing apps
   onto Apple Watch

» Discovering settings in the
   iPhone's Apple Watch app

» Using Apple Watch as an extra
   security tool for Mac owners

» Enabling Apple Watch's
   accessibility settings, including
   AssistiveTouch

» Playing around with
   recommended third-party apps

Chapter **11**

App It Up: Customizing Apple Watch with Awesome Apps and More

N o two people are exactly alike, so why should our devices have the same apps installed?

Just like it's fun to customize a smartphone or tablet screen with all kinds of hand-picked apps — short for "applications," which is just a more trendy way to say "software" or "programs" — Apple Watch supports many thousands of third-party (non-Apple) apps.

The App Store for Apple Watch is already jammed with apps designed for your sleek wearable, including many familiar apps (perhaps based on its iPhone or iPad brethren) and completely new apps looking to dominate this new screen in your life. In most cases, they're free to download (or close to it), and you're sure to find something that resonates with you.

Bottom line: We're all individuals who have different tastes, priorities, and interests.

You can now download apps directly onto Apple Watch, instead of requiring an iPhone.

This chapter looks at how to choose and install Apple Watch apps, and manage them from your Home screen. I take you through the companion Apple Watch app for iPhone and how to tweak settings and personalization options. Toward the end of this chapter, I also suggest several stellar apps you might want to consider for your watch. (See Chapter 12 for a brief discussion about gaming apps for your Apple Watch.)

Downloading Apps for Apple Watch

Because of the popularity of Apple's App Store, perhaps it's no surprise developers want to capitalize on watchOS and the Apple Watch. After all, millions of customers are wearing this gadget on their wrists and will likely want to customize their experience with apps.

In case you weren't aware, developers make 70 percent on paid apps, with the remaining 30 percent going to Apple. Thus, an opportunity exists to make some serious cash by creating a must-have app for Apple Watch.

"The more you wear Apple Watch, the more you'll realize just how personal a device it is," says Apple on its website. "Because with so many different apps available, you can choose the ones that are most relevant to you and create a customized experience.

"There are already apps for airlines, department stores, social networks, and more that take advantage of the unique opportunities the wrist brings. And with new apps being built for Apple Watch every day, this is just the beginning," adds Apple.

How does one go about downloading apps?

From the App Store, but not the same App Store as the one on your iPhone, iPad, iPod touch, or Mac. It's a separate store dedicated solely to Apple Watch, as shown in Figure 11-1.

FIGURE 11-1:
A look at the App Store for Apple Watch, which is part of the Apple Watch app on iPhone.

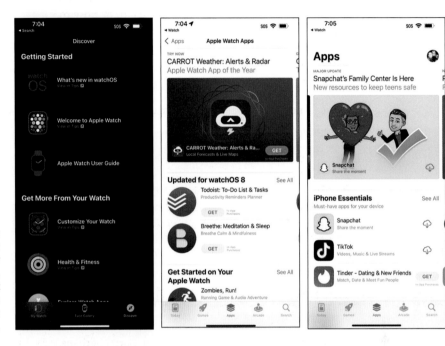

As previously mentioned, you can download apps to Apple Watch in two different ways:

» Use the Apple Watch App Store built into the Apple Watch app on your iPhone.

» Download apps directly to Apple Watch (requires watchOS 6 or newer).

From your iPhone

To use the Apple Watch App Store on your iPhone, follow these steps:

1. **On your iPhone, tap the Apple Watch app.**

   The black icon with a silver Apple Watch is preinstalled on all iPhones. Tapping the app launches the App Store for Apple Watch.

2. **Tap the Discover tab along the bottom of the screen. Then tap Explore Watch Apps, at the bottom of the screen.**

   You see some recommended apps to get you going.

3. **Find apps to download based on what's featured or by category.**

   For example, you can tap the See All tab beside Get Started to open a list of popular apps. Alternatively, tap the Search tab (as shown in Figure 11-2) in the lower right of the iPhone screen if you'd prefer to look for something by keyword instead of browsing.

   Other tabs at the bottom of the screen include: Today (a collection of apps highlighted each day), Games, Apps, and Arcade (which requires a subscription to the Apple Arcade gaming service).

   Note: Despite the fact the App Store is accessible inside the Apple Watch app on iPhone, be aware not all these apps work with Apple Watch (some are for iPhone only). But you see which ones are supported before you download.

FIGURE 11-2:
Use keywords, such as "apple watch games," to find new content for your Apple Watch.

Directly to your Apple Watch

To use the Apple Watch App Store on your Apple Watch, follow these steps:

1. **Tap the App Store icon on Apple Watch to launch the store.**

2. **If you haven't done so already, sign into the App Store. Tap Password and use the virtual keyboard (on a nearby iPhone) or Scribble feature to type it (only required once).**

   You might also be prompted to add a passcode to your Apple Watch to download apps, if you haven't already set that up.

3. **Browse through all the content to find something to download and try.**

   Search using Scribble or Dictation, or rely on your Siri voice assistant. Read reviews, look at the overall star rating, and check out screen grabs.

After it downloads, the app is available on your Apple Watch Home screen. You may be prompted first with a message like "Double click [side button] to install" or "Requires iPhone to complete setup." Just follow along! See Figure 11-3.

FIGURE 11-3:
You can download Apple Watch apps directly to your wrist. Here's what they look like at the mini App Store.

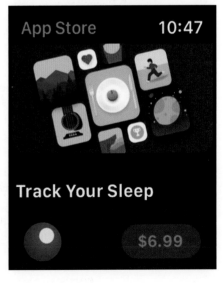

Many of the Apple Watch apps are free, and some require a nearby iPhone to get the most from them. Just remember that most Apple Watch apps have a companion app for iPhone; therefore, you need to have enough storage on your iPhone — not just on the watch — for many of them.

While at the App Store on your Apple Watch or iPhone, it's a good idea to read the app's full description and check out the screenshots before you download it to ensure it's for you before you waste your time, data, and, perhaps, money. Also, read reviews by those who've already used the app.

You can also update the Apple Watch's operating system directly on the watch. Tap the Settings app icon, followed by Software Update, and if there's a newer operating system to download, it prompts you to download the update.

As most Apple Watch apps also install an iPhone version, after you download it, it syncs between devices. In other words, you don't need to do anything to copy apps over to the watch (if you downloaded Apple Watch apps to iPhone) or vice versa (if you downloaded to Apple Watch directly). The moment the Apple Watch and iPhone are synced over Wi-Fi or Bluetooth, whatever you downloaded is transferred to your wrist.

Deleting apps

To manage the apps on your Apple Watch — to delete ones you no longer want, for example, or perhaps to tweak the layout of them — you can do this in one of two ways.

>> **Use the Apple Watch app on your iPhone.** Simply launch the Apple Watch app on your iPhone and then tap My Watch, where you can uninstall apps you no longer want on your Apple Watch.

>> **Delete apps directly off Apple Watch.** From your Apple Watch's main Home screen — where you see all the circular icons in "Grid" view — simply press and hold lightly on an icon and they all start jiggling. If the app has a little "X" in the corner (see Figure 11-4), you can tap it to delete it off Apple Watch.

If the app doesn't have a little "X," which is most (but not all) of the built-in Apple Watch apps, you can't delete them.

Unless you have a generous data plan, only download large apps on your iPhone when you're in a Wi-Fi hotspot. A couple megabytes is fine if you're using cellular data, but you should use wireless broadband for anything larger than that. Some games can be a couple gigabytes in size (1 gigabyte is roughly 1,000 megabytes).

FIGURE 11-4:
You can delete third-party apps from Apple Watch itself. Tap the little "X" to remove.

Adjusting Settings in the Companion Apple Watch App on Your iPhone

You have tons of settings to change on your Apple Watch app. From changing the orientation of your watch to match your dominant hand to changing how your watch looks and sounds, the following sections cover all the details that help you make your watch yours. In this section, you also see personalization, general, accessibility, and first-party app options.

Settings and personalization

Instead of tapping the App Store tab in the lower right of the Apple Watch app on iPhone where you download Apple Watch apps, tap the My Watch tab in the lower left to open the settings and personalization options for the watch.

From within My Watch, you can perform the following actions:

» Change the orange **App View** of the apps on your Apple Watch's Home screen (Grid View, which is on by default, or List View).

>> Near the top of the screen, where you can see what **Apple Watch** you have (such as 44mm Case - Aluminum), you can **unpair** your Apple Watch from your iPhone.

>> You see the various **Watch/Clock Faces** illustrated at the top of the screen. Tap to select one.

>> Select **Complications** to change what info you see on your favorite clock faces.

>> The **Notifications** tab lets you customize what notifications you want pushed to the watch.

>> Select **Dock** to see (and customize) what apps you want to see when you press the side button.

>> Adjust brightness and text size for the watch in the blue-and-white **Brightness & Text Size** section. As shown in Figure 11-5, use the slider to adjust each setting and choose to bold text or not.

>> Enable or disable sounds and vibrations in the red-and-white **Sounds & Haptics** area (also shown in Figure 11-5). You should have options for Alert Volume, Mute, Cover to Mute (covering your watch with your hand to mute it), Haptic Strength, and Prominent Haptic for more pronounced vibrations.

>> The purple **Do Not Disturb** tab turns off all alerts, sounds, and haptics to the watch, whereas the orange **Airplane Mode** disables the wireless connection between Apple Watch and the iPhone. When enabled, the **Mirror iPhone** option automatically turns off Apple Watch's wireless radios whenever you do the same on the iPhone.

>> You should also find options for **Accessibility, Siri, Passcode, Emergency SOS, and Privacy.**

Have fun exploring what you can customize on Apple Watch. There's quite a bit.

General options

Under General options for Apple Watch, you should find the following information:

>> **About:** Software version installed and other information.

>> **Software Update:** To look for and download the latest watchOS update.

>> **Automatic App Install:** For watchOS and app updates.

>> **Watch Orientation:** Left or right wrist and whether the Digital Crown button is on the left or right side. You have the ability to change this, as shown in Figure 11-6.

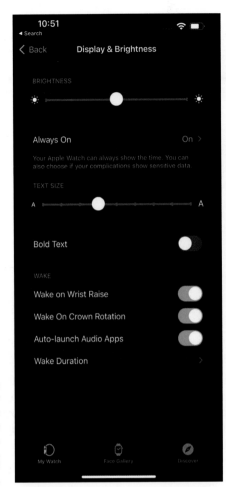

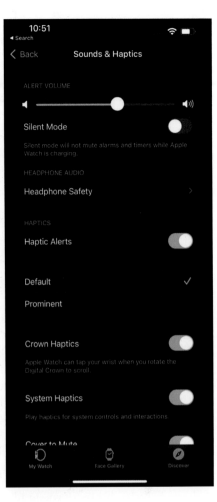

FIGURE 11-5:
Use your
fingertip to
adjust settings
for brightness,
text size,
sounds, and
haptics.

>> **Accessibility:** Options/aids for seeing and hearing impaired and more.

>> **Language & Region:** Choose the System Languages, Region, and Calendar type (which by default is Gregorian).

>> **Apple ID:** The account you're signed in to.

>> **Enable Handoff:** When this is on, your iPhone picks up where you left off with supported apps on your Apple Watch.

>> **Wrist Detection:** When this is on, Apple Watch automatically shows you the time and latest alerts when you raise your wrist.

Other settings are here, too, such as Wake Screen (how and when to wake the Apple Watch screen), Enable Dictation (to enable on-device dictation), Usage, Diagnostic Logs, and more.

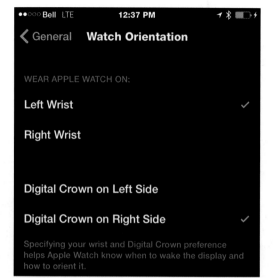

FIGURE 11-6:
You can
change the
watch's
orientation
from left wrist
(default) to
right and
which side the
Digital Crown
button is on.

Accessibility options

Those living with visual, aural, or mobility/dexterity-related impairments can take advantage of a number of accessibility options on Apple Watch. Just as Apple has done with its iOS, iPadOS, and OS X products (and the Safari web browser), Apple's smartwatch provides several features you can enable in the Accessibility area of the Apple Watch app on your iPhone, as shown in Figure 11-7. I cover Apple's vast (and growing) accessibility features in Chapter 2, plus additional details are available at Apple's website for these options (www.apple.com/accessibility/watch).

For those who'd like a quick summary, however, here are just a few of the aids you can enable:

>> Visually impaired users can enable **VoiceOver,** a screen reader available in one of several languages.

>> Enlarge the Apple Watch fonts in **Font Adjustment** (ideal for the Mail, Messages, and Settings apps), add **Bold Text,** and opt for the **Extra Large Watch Face** option with bigger numbers to read.

>> Other visual tweaks include **Zoom** (magnification), **Grayscale** (no color), **Reduce Transparency** (increases contrast), **On/Off Labels** (easier to see if setting is on or off), and **Reduce Motion** (the Home screen is easier to navigate).

>> For those with impaired hearing, Apple Watch lets you enable **Mono Audio** (sending both audio channels to both ears and letting you adjust balance for greater volume in one ear or the other).

>> Apple Watch's Taptic Engine can be enhanced with **Prominent Haptic,** a setting that gives a slightly more noticeable vibration and "preannouncement" for some alerts, such as calendar appointments, messages, phone calls, and more. A *preannouncement* means the watch sends you an additional haptic vibration before the main one.

>> There is also a **Tap to Talk** feature (Walkie-Talkie app), **Air Pods** tap options, **Chimes** (when to hear chimes per app), **Side Button Click Speed,** and **Touch Accommodations** (on or off).

>> For those with dexterity challenges, **AssistiveTouch** helps you use Apple Watch if you have difficulty touching the screen or pressing the buttons. The built-in sensors on Apple Watch can help you answer calls, control an onscreen pointer, and launch a menu of actions -- all through hand gestures.

That is, AssistiveTouch lets you navigate your Apple Watch with one hand: the same as the one wearing Apple Watch.

Here are the three main moves:

- *Move to the next item:* Pinch (tap your pointer finger to your thumb).

- *Move back one item:* Double-pinch (tap your pointer finger to your thumb twice quickly).

- *Tap an item:* Clench (Close your hand into a fist) is the equivalent to a "tap." See Figure 11-8.

- *Show the Action Menu:* Double-clench (two quick open and closing moves with your first).

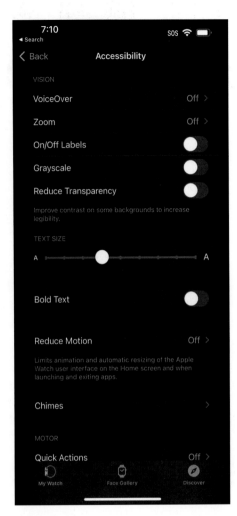

FIGURE 11-7:
You can adjust a number of accessibility options for Apple Watch, as shown in the Apple Watch app for iPhone.

FIGURE 11-8:
When
AssistiveTouch
is enabled,
one of the
moves is to tap
an on-screen
item, handle
via clenching
your fist.

Let's learn more about AssistiveTouch, and how to use it, as it really is fascinating and could be a game-changer for many.

Using gestures with AssistiveTouch, you can tap the screen, press and turn the Digital Crown, swipe between screens, hold the side button, show all your apps, use Apple Pay, activate Siri, and access the Notification Center, Control Center, and the Dock.

To enable AssistiveTouch, start with the Apple Watch app on your iPhone:

1. **Go to Accessibility ⇨ Motor ⇨ AssistiveTouch, then turn on AssistiveTouch.**

2. **Now tap Hand Gestures, then turn on Hand Gestures. Review the four main gestures and what they do.**

   You can tap where it says "Learn More" to, yep, learn more about these moves (and you can follow along with the animation).

   There is also an option for a Motion Pointer, which you can enable here in the Accessibility settings (under Motor). In addition to pinching and clenching, with the Motion Pointer you can control your Apple Watch by tilting the watch up and down and side to side.

Although you can also see and enable accessibility features on the Apple Watch itself, it's probably easier to explore all these helpful options in the Apple Watch app on iPhone. See Figure 11-9.

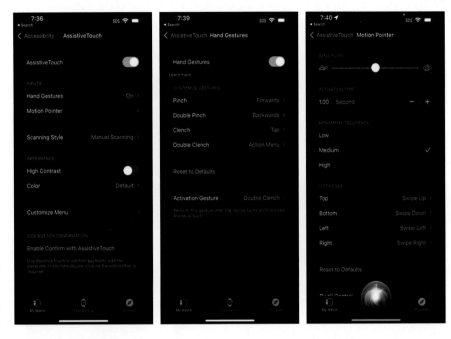

FIGURE 11-9:
Some of the
AssistiveTouch
options for
gesture
control.

REMEMBER

You can also adjust some accessibility options on Apple Watch itself by triple-tapping the Digital Crown button.

New for 2022, a feature called Apple Watch Mirroring makes the watch experience more accessible for people with physical and motor disabilities. How? When enabled — in the Accessiblity options of the iPhone's Apple Watch app — Apple Watch Mirroring allows you to stream your Apple Watch to your iPhone and fully control it using assistive features on iPhone like Switch Control and Voice Control. See Figure 11-10.

Apple Watch Mirroring is part of Apple's AssistiveTouch features.

To use Apple Watch Mirroring, be sure you're running at least iOS 16 and watchOS 9. Then

1. **Open the Settings app on your iPhone.**

2. **Tap Accessibility. Apple Watch mirroring in Accessibility settings.**

3. **Tap Apple Watch Mirroring.**

4. **Now tap the green toggle button to enable Apple Watch mirroring.**

   After a moment or two, a new screen appears on your iPhone. This pop-up box on your iPhone will essentially be using AirPlay between your iPhone and Apple Watch, so you have an easier way to interact with your Apple Watch.

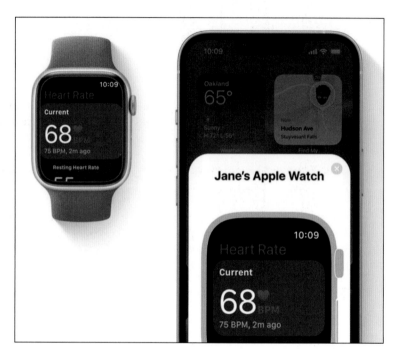

FIGURE 11-10:
If the small Apple Watch screen is difficult to interact with, a new accessibility feature called Apple Watch Mirroring lets you see a virtual version of Apple Watch on your iPhone.

Security options

Beginning with watchOS 6, Mac users have the ability to leverage their Apple Watch as an extra security tool.

As shown in Figure 11-11, the Mac includes an option to approve security prompts by tapping on the side button of Apple Watch, when prompted, to confirm it's really you logging in! Enable this option via the Mac's System Preferences ⇨ Security & Privacy area.

You can also use Apple Watch to display Apple ID verification codes, for when you need to log in to your Apple account on a new device or browser ("two factor authentication" combines something you know, such as a password, with something you have, such as Apple Watch).

FIGURE 11-11:
Your Apple
Watch can help
log you into a
Mac or display
a one-time-use
code to log in
somewhere.

You can also have your Apple Watch automatically verify you to unlock your Mac and apps. Make sure your Mac has Wi-Fi and Bluetooth turned on and then sign into iCloud with the same Apple ID as your Watch.

Note: You must have two-factor authentication enabled on your Mac, and your Apple Watch is using a passcode.

On your Mac, choose Apple menu ⇨ System Preferences. Now click Security & Privacy. Select "Use your Apple Watch to unlock apps and your Mac" or "Allow your Apple Watch to unlock your Mac."

First-party app options

Of course, multiple options and settings can be explored in each of the first-party apps installed on Apple Watch, such as: Activity, Calendar, Clock, Contacts, Health, Mail, Maps, Messages, Music, Phone, Photos, Reminders, Sleep, Stocks, Weather, Wallet & Apple Pay, and Workout. Some examples of what you can change include:

» **Clock:** You can select 12- or 24-hour (military) time; Push Alerts from iPhone (if enabled, Apple Watch lets you know about iPhone alerts); Notifications Indication (if enabled, a red dot appears at the top of the watch face when you have unread notifications); Monogram (choose a one- to four-letter monogram to appear as a monogram complication on the Color watch face); and City Abbreviations (specifying different acronyms for the default ones provided for your chosen cities for the World Clock complications).

» **Activity:** You can choose to turn on or off Stand Reminders (notified 50 minutes into an hour of inactivity); Progress Updates (set a time interval to receive an update on your Activity progress, such as every five hours); Goal Completions (on or off); Achievements (on or off); and Weekly Summary (on or off — received every Monday).

- » **Messages:** You can see all the default replies or change them to anything you like.

- » **Music:** You can select Synced Playlist (this playlist syncs when Apple Watch is on its charger) and Playlist Limit (such as changing it from 2 gigabytes to 1 gigabyte).

Twenty Recommended Third-Party Apple Watch Apps

Chapter 3 looks at all the default apps preinstalled on Apple Watch — you know, the ones Apple put there and you can't remove (which understandably frustrates a number of people) — but I want to share a number of optional but recommended and free third-party apps you can download from the Apple Watch App Store. All of them take advantage of Apple Watch's features. Because I reserve games for Chapter 12, these 20 apps are tied more to information, travel, automotive, productivity, social media, fitness, shopping, and some entertainment.

- » **Mint:** The Apple Watch app for this popular finance tool lets you view your monthly spending goals at a glance as well as track your progress toward meeting them. And for those trying to stick to a budget, you can choose to receive weekly alerts with insight on how well you're doing (or not).

- » **ESPN:** One of the most popular sports apps for iPhone is available for Apple Watch. Select which sports matter to you — such as baseball, football, basketball, hockey, golf, or tennis (or all of these) — and stay up to date with breaking sports news, real-time scores, and more.

- » **Target:** As one of the first retailers to support Apple Watch, Target has an app that lets you build and view a shopping list on your watch so you can glance down to see what items you need — even if your phone is tucked away in your purse or pocket. When you enter a store, the Target app also tells you where to find the items you're looking for.

- » **OneDrive:** Although once bitter rivals, Microsoft has embraced Apple's iOS platform — and now the watchOS, too. Based on Microsoft's OneDrive cloud service, this Apple Watch app lets users see their stored photos on their wrist — even when an iPhone isn't nearby.

- » **Marriott Bonvoy:** One of the cooler apps is Marriott Bonvoy, which lets you unlock your hotel door by waving your Apple Watch at the sensor. A room key isn't required. The official app can also provide directions to your hotel, check you in, show your points balance, and more.

- » **Chirp for Twitter:** While Twitter mysteriously pulled its Apple Watch support in 2017, you can see your Twitter feeds right on your wrist — with the help of a third-party app, Chirp for Twitter. And because they're only a couple of hundred characters, tweets fit perfectly on Apple Watch's small screen. Feel a gentle tap whenever new tweets are posted; plus you can retweet and favorite tweets from your Apple Watch. You can also send and receive direct messages, see the latest trending topics, or search by hashtag.

- » **OpenTable:** Hungry? The OpenTable app supports Apple Watch, which lets you see information about your upcoming dinner reservations by simply looking down at your wrist. The app can also help guide you to the restaurant with turn-by-turn directions.

- » **Evernote:** A popular productivity tool, Evernote for Apple Watch lets you view your stored notes, dictate a new one, set reminders, and search by keyword if you're looking for something in particular. Because Evernote stores your notes in the cloud, you can view your dictated notes in other Evernote apps — perhaps on a smartphone, tablet, or laptop.

- » **American Airlines:** How do you know when it's time to leave for the airport? Or if your flight has been delayed, cancelled, or changed gates? American Airlines (AA) has an Apple Watch app that can alert you to any and all of these things. The AA app also lets you check in for your flight, view a map with your estimated time of arrival, view baggage claim and connection details, and more.

- » **BMW i Remote:** Own an electric BMW i vehicle? The official Apple Watch app lets you remotely check on the charge status or notifies you when your car has been fully charged and is ready to go. This smartwatch app also lets you check your miles (to prevent "range anxiety"), see door-lock status, get service reminders, and view your cabin temperature.

- » **CNN:** News junkies, rejoice! The official CNN app for Apple Watch gives you the information you need wherever life takes you. Select to receive breaking news and developing stories based on 12 categories of interest — such as Top Stories, U.S., World, Politics, Health, Entertainment, Sports, and Technology — plus your watch can even launch CNN TV live on your iPhone.

- » **eBay:** The world's largest marketplace is now a tap away. eBay on Apple Watch helps you keep up with the auctions you're watching — whether you're bidding on something or selling merchandise. The app conveniently lets you send and receive alerts without having to fumble through your phone, tablet, or personal computer.

- » **Citymapper:** If you rely on public transit, the Citymapper app for Apple Watch always shows you the best bus and train routes based on your location and where you want to go. You should see step-by-step instructions, including a list of the next three arrival times for your mode of transportation so you can decide when to leave, and you should feel a vibration on your wrist when it's time to get off at your stop.

» **TripAdvisor:** Find that hidden gem of a restaurant on your next trip. Unearth dozens of things to do while discovering a new town. Everything that makes TripAdvisor the perfect travel companion is now on your Apple Watch. Get instant information on hundreds of nearby restaurants, sights, and tourist destinations.

» **NPR One:** Fans of NPR can make their favorite station even more personal. The NPR One app shows you relevant news and curated stories based on your interests, along with access to your playlist (on your iPhone), and you can search for specific shows by using dictation and control basic playback functions with your fingertip.

» **Fandango:** The popular movie ticketing app is now on your wrist. After you've purchased tickets to a flick, the Apple Watch app can display the movie time and theater location, phone number, and other information you might need.

» **Things:** If you wear a watch — and a smartwatch, no less — it might not be a stretch to assume you like to be organized. But that doesn't mean you're good at it! If you need a little help, an app called Things is an excellent to-do manager for iPhone, and it supports Apple Watch too. Organize your life with daily tasks, which you can easily sort into sections like Today, Upcoming, and Anytime. Specifically, the Apple Watch app focuses on your current tasks, which can be displayed as a watch face complication or in the app itself, and lets you tick off items when completed.

» **PayByPhone Parking:** You can use the PayByPhone Parking app on Apple Watch to pay the meter, check on the time remaining, and deliver an alert ten minutes before the meter expires. If you're not done with your errands, your watch lets you add more time to the meter without your having to go back to your car.

» **Sky Guide:** Watch the skies! And your wrist. The Sky Guide app for Apple Watch is great for armchair astronomers. Receive alerts about upcoming celestial events — such as meteor showers and eclipses — and it even alerts you when the International Space Station is about to fly over your location.

» **Lutron Caséta:** Your smartwatch can control your smart home. The Lutron Caséta app for Apple Watch lets you control the lights in your home even when you're not there so you can make it look like you're home when you're on vacation. Or on the flipside, if you accidentally leave the lights on when you leave, you can get an alert on your wrist to turn them off.

Chapter **12**

Extra! Extra! Having Fun with Apple Watch

Y ou didn't think Apple Watch was just for information, communication, and navigation, did you?

This mobile companion of yours is also ideal for playing around. In Chapter 5, I look at how to send animated emojis and digital sketches to people. And Chapter 9 covers music and podcast playback, audiobooks, and how to control Apple TV playlists from your wrist.

But your watch can do much more in the fun department.

Granted, Apple doesn't seem to advertise these nonessential applications as much as customizing watch faces, sending messages, or calculating your physical activity, but you can indeed enjoy some downtime with Apple Watch, including many playable games already available in the App Store for Apple Watch.

You can also look at photos of people, pets, and places on your Apple Watch — anytime and anywhere. Perhaps you ran into someone who asked how old your daughter is. Now you can show that person her smiling face. Or maybe you want to glance down to see old friends when a song on the radio brings back camp memories. If you're feeling like you need a vacation, call up photos of last year's trip to Jamaica so you can see the white sand and blue water (and then call your travel agent!).

Speaking of photos, Apple Watch can let you access your iPhone's camera to snap the shutter button wirelessly, which is ideal for selfies and group shots. This is available through the Camera app.

Copying Photos to Apple Watch

If you own an Apple Watch without cellular access, you likely know most tasks require a nearby iPhone. The two devices are wirelessly tethered via Bluetooth and Wi-Fi, but you can sync some files to Apple Watch just in case your iPhone isn't nearby.

Chapter 9 talks about syncing music to the watch — perhaps if you want to go for a jog around the neighborhood without your phone and you own a Bluetooth headset — but here, I cover transferring photos over to your watch.

I'm not exactly sure *why* you'd want to copy photos to your watch, unless you really think you're going to want to see some pictures when your iPhone isn't nearby. Having music on the watch makes more sense — per my jogging scenario — but those who want to take advantage of this feature must first enable it on the Apple Watch app on iPhone.

To copy photos onto your Apple Watch, follow these steps:

1. **Open the Apple Watch app on your iPhone and tap My Watch in the bottom left of the screen.**

2. **Swipe down until you see the Photos icon and then tap it.**

3. **Some options appear, including one to mirror your iPhone (where photos are synced), or you can choose Custom to handpick what's synced to the watch.**

As shown in Figures 12-1 and 12-2, you've got some options in the Photos area of the Apple Watch app on iPhone:

- *Synced Album:* Select which iPhone-stored photos are viewable on Apple Watch, even when you don't have your iPhone with you. By default, it's your Favorites album, but you can also choose another one, such as Camera Roll. Or select None.

- *Photos Limit:* Select the photo storage limit on your Apple Watch. Although this limit is measured in megabytes (MB), you can raise or lower the number, but 15MB is the default number, which translates to 100 photos. Lowering it to 5MB loads only 25 photos. You can raise it to 40MB (250 photos) or a maximum of 75MB (500 photos).

Be aware you can also view photos from your iCloud account on Apple Watch.

REMEMBER

Interestingly, you can also copy up to 2 gigabytes of music to Apple Watch, which translates to roughly 500 songs. Therefore, you can store up to 500 photos and up to 500 tunes. You can't go over this maximum for photos and music.

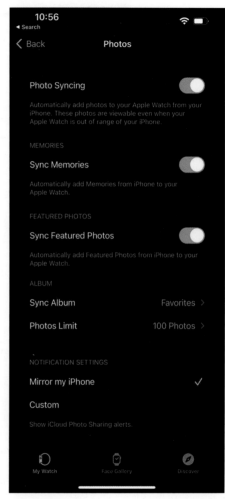

FIGURE 12-1:
Select which photos you want synced to Apple Watch (if any).

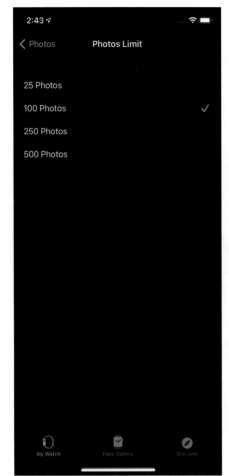
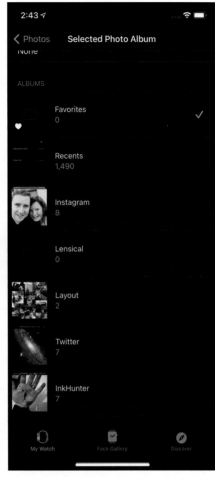

FIGURE 12-2:
Select how many photos you'd like to sync (by size or number of files) and from which album.

Launching Photos on Apple Watch

Okay, so Apple Watch doesn't have the biggest screen in your life, but it *is* always on your wrist; therefore, it's a conveniently placed digital photo frame.

One of the built-in apps is Photos, which is similar to a photo gallery app on your iPhone or iPad.

To use the Photos app on your Apple Watch, follow these steps:

1. **Tap the Digital Crown button to go to the Home screen.**

2. **Tap the Photos app.**

   If you prefer, raise your wrist and say "Hey Siri, Photos." Either action launches the Photos app, which shows you thumbnails of photos stored on your phone (or watch).

3. **Twist the Digital Crown button to zoom in and out on individual images.**

   Twisting the Digital Crown away from you zooms out to see more photos, which makes the thumbnails smaller, whereas twisting toward you zooms into a photo.

4. **Zoom in until a photo takes up the entire watch face.**

   Now you can swipe left and right to browse through your photos one at a time. The photos are in the same order as they are on your iPhone's Photos app, including any albums you've created. Figure 12-3 shows a photo full screen on Apple Watch.

   Why doesn't Apple Watch allow you to pinch and zoom? Your finger and thumb would cover up your photos. The Digital Crown button works better.

FIGURE 12-3:
Swipe your finger left or right to scroll through all your photos. Hey, that's the author of *Apple Watch For Dummies* and his better half (on roller skates, no less)!

5. **Twist the Digital Crown button to zoom out to see more photos, or tap the Digital Crown button to exit the Photos app and return to the Home screen.**

   Speaking of photos, don't forget Apple Watch can show you images that are embedded/attached to Messages and Mail (email). You should see images just below the text in a given message. Twist the Digital Crown button to see the accompanying photo(s) near the bottom of the screen. Too bad you can't send a photo to someone from Apple Watch. For that, you need to use your iPhone.

Choosing a Photo for Your Portraits Watch Face

As you likely know, Portrait mode on iPhone is a fun photo feature as it intelligently enhances the person's face and blurs out the background. Beginning with watchOS 8, your Apple Watch can now take advantage of this feature, as part of its Portraits watch face.

Essentially, it uses depth data to crop out the person in the photo and adds a slightly animated look to the still photo, therefore tilting your wrist makes the image tilt accordingly.

On Apple Watch, you can store a total of 24 photos in the Portraits watch face and shuffle between them or zoom in to enjoy the 3D effect.

Here's how to set Portrait mode photos from your iPhone:

1. **Launch the Watch app on your iPhone.**

2. **Open the Face Gallery tab.**

3. **Scroll down and tap the Portraits option.**

4. **Tap Choose Photos under the Content section.**

   Here, select the photos you want as the watch face and tap Add.

5. **Choose a photo to set Portraits watch face on Apple Watch.**

   Choose a Style, such as Classic, Modern, or Rounded.

6. **In the Complications section, select Middle (Date) and Bottom to select suitable complications, or turn them off for a clean effect.**

7. **Tap Add next to the Portraits.**

Now your Apple Watch will sport the new Portraits watch face. To check out all the photos, lightly tap the watch face. You may notice the watch automatically moves the time and other complications according to the image, but you can manage this yourself if you want to customize it. See Figure 12-4.

TIP

You can zoom in or out of the Portrait photo by rotating the Digital Crown.

Beginning with watchOS 9 in 2022, the Portraits watch face now lets you put pictures of your dog or cat on it, too! And you can customize your image even further by using the editing mode to add a tint to the background layers of a photo. See Figure 12-5.

FIGURE 12-4:
The Portraits watch face on Apple Watch uses Portrait mode photos from your iPhone to create a multilayered watch face with depth.

FIGURE 12-5:
Now supporting pets! The Portraits face for Apple Watch isn't just for humans anymore! Chapter 4 goes into great detail about customizing your clock faces.

To change the Style or Complications:

1. **When on the watch face, press and hold on the screen.**

2. **Tap Edit. Next, look through the options and select a suitable one.**

TIP

You can also share your new Portraits watch face! From the edit screen, tap on the Share icon, select the contact, and choose to send via iMessage or Mail.

To delete Portraits faces from your Apple Watch, go to the Watch app on iPhone and select the ones to delete in the My Faces section.

Beginning with watchOS 8, photo highlights from your Memories and Featured Photos now automatically sync to your watch, serving up something new every day. "On This Day" photos (shown in Figure 12-6) are displayed in a unique mosaic grid, highlighting memories captured on that exact day.

FIGURE 12-6:
Just like your featured Memories are displayed on your iPhone, those same photos are featured on your Apple Watch, in a mosaic view.

Discovering the Camera App

Did you know you can use Apple Watch as a viewfinder for your iPhone's rear-facing camera? You can see a preview of your photo before you take it, set the camera timer on your watch, or just take the photo.

The Camera app might be ideal for those who want to get in the picture but don't want to press the iPhone's shutter button (on the screen or a button along the side of the phone). It can be awkward to hold the camera and take a selfie at the same time or, worse, risk dropping your iPhone and breaking the glass. Or you can use one of those selfie sticks too.

To use the Camera app on your Apple Watch, follow these steps:

1. **Tap the Digital Crown button to go to the Home screen.**

2. **Tap the Camera app.**

   Alternatively, you can lift your wrist and say "Hey Siri, Camera." Either action launches the Camera app. You should then see a preview of your iPhone's camera.

3. **Frame up your shot by getting your subjects huddled together or centering a landscape photo or whatever, as shown in Figure 12-7.**

   Consider Apple Watch your live viewfinder for the iPhone you're holding (or one you've placed on a tripod).

FIGURE 12-7:
No broccoli in the teeth? As you can see here, you can take a selfie or a group shot with friends.

4. **If you want to take a photo, press the white shutter button in the center of the watch — just below the preview window.**

   This instantly snaps the picture and adds a thumbnail to the bottom left of your Apple Watch's screen. You can tap this thumbnail if you'd like to see the photo full screen.

5. **If you want to use the timer, tap where it says 3s (3 seconds) and then tap the white shutter button (shown in Figure 12-8).**

    You see a countdown on the screen before the photo is taken.

    Review what you took, and if you like it, press the Digital Crown button to return to your Home screen. Don't bother pressing the white shutter button if you don't like what you see. Frame up a different shot and then press the white button.

FIGURE 12-8:
Whether you use the timer or not, tap the large white shutter button to snap a picture by using your iPhone's camera. It counts down from 3 to 1.

Here is a quick checklist of what you can do when viewing photos on the Camera app on Apple Watch.

First, to view a photo: Tap the thumbnail in the bottom left. Then:

>> **See other photos:** Swipe left or right.

>> **Zoom:** Turn the Digital Crown.

>> **Pan:** Drag on a zoomed photo.

>> **Fill the screen:** Double-tap the screen.

>> **Show or hide the Close button and the shot count:** Tap the screen. When finished, tap Close.

Or, when not viewing a photo, tap the three dots in the lower-right of the Camera app on Apple Watch, to access:

>> **Timer:** (3 seconds or none).

>> **Camera:** (front or rear, as shown in Figure 12-9).

>> **Flash** (auto, on, or off).

>> **Live Photo** (auto, on, or off): If your iPhone supports it, the camera records what happens 1.5 seconds before and after you take a picture.

>> **HDR** (on or off): If your iPhone supports it, high-dynamic range results in better contrast, brightness, and color.

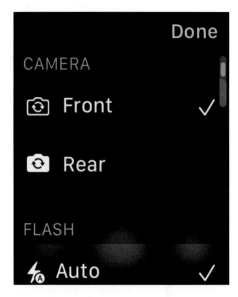

FIGURE 12-9:
Scrolling through the photography options on the Camera app on Apple Watch.

Examining a Batch of Apple Watch Games

When smartphones started to take off in the early part of the 21st century — including iPhone's high-profile debut in 2007 — many video game purists said the screen was simply too small to provide a gratifying interactive entertainment experience. Why would people want to squint to play a game on a (then) 3.5-inch screen when they have a 14-inch laptop, 23-inch desktop, and 60-inch flat-panel TV at home? And don't you need a game controller with real buttons as opposed to finger taps, swipes, and flicks on a small screen?

But then along came a handful of stellar games that proved the naysayers wrong: Angry Birds, Temple Run, Flappy Bird, Subway Surfers, Cut the Rope, Candy Crush Saga, Fruit Ninja, Words With Friends, and Jetpack Joyride, to name just a few.

Gaming on a smartphone is not only convenient — because we don't go anywhere without these devices — but the App Store offers a ton of games, with most downloadable games costing just 99 cents, if anything at all.

Whether Apple Watch will be a gaming platform that withstands the test of time is still up for debate, and developers have a few obstacles to overcome — a tiny screen, limited user interface, and questionable battery life — but where there's a will, there's a way — and an app for that! Gamers love to play anywhere, anytime, and on any device, so "time" will tell whether the wrist will be the new place to game.

That hasn't stopped a number of game developers from creating digital diversions on Apple Watch, many of which were announced even before the watch became available in spring 2015.

The following is a look at ten great games. Remember, to find and download these games and other apps, you can do it in two ways: Open the App Store on your Apple Watch or launch the Apple Watch app on your iPhone and then tap the App Store tab in the lower-right corner of the screen.

Lifeline 2

Even if you never played the original Lifeline — for iPhone, iPod touch, and iPad — the 99-cent sequel, which is also playable on Apple Watch, proves to be a challenging text adventure for gamers on the go.

In fact, Big Fish Premium's Lifeline 2 offers a new story almost twice as long as the original, with many more choices and paths to explore. Playing out in real-time, the story surrounds Arika, a young woman on a deadly quest to avenge her parents and rescue her long-lost brother.

Along with strong writing and tough decisions, Lifeline 2 features a 28-minute original soundtrack to help you stay immersed in the adventure.

Trivia Crack

One of the biggest trivia games on iPhone is also available for Apple Watch. Trivia Crack lets you play rounds of pop culture trivia — without ever taking your iPhone out of your pocket. Animated characters guide you in defeating all your opponents while you answer millions of questions submitted by users from around the world. Questions appear on the watch's screen, and you can answer them right on the watch; after an opponent answers and it's your turn, you're notified from your watch and can continue playing from there. Example: What's Muhammad Ali's real name? Rocky Balboa, Rocky Marciano, Anderson Silva, or Cassius Clay (correct answer).

Pocket Bandit

Have a spare 99 cents? It's a "steal" for a great game like Pocket Bandit. It's a fun and fast-paced puzzler that offers more than 100 treasures to find (including seasonal updates, such as the ability to steal the Santa Claus factory around the holidays). From developer Seele, this attractive game has you use the Digital Crown to move the cursor. When the Apple Watch starts vibrating, it means that you are on a good number. Touch the Apple Watch to validate the number. If it's a good number, the safe opens and you win a treasure.

Best Fiends

If you like such matching games as Candy Crush Saga, Apple Watch owners might enjoy tapping through Best Fiends — one of the first mobile games for the platform and based on the popular iOS version. In the single-player puzzle game from developer Seriously, players explore the lands of Minutia, collect treasures, and battle the malevolent Slugs of Mount Boom. What's more, players earn rewards that can be applied to their Best Fiends game on their iPhone — used toward leveling up characters, unlocking new powers, and defeating enemies. In the iOS version, players drag a finger up, down, and diagonally to match identical items on an obstacle-laden game board.

Snappy Word

Australian developer Right Pedal has released Snappy Word, a word game that places four random letter tiles on the watch face. By tapping the letters in order to create words, your goal is to see how many words you can create in 30 seconds. For example, E, M, S, and A can create such words as *same*, *mesa*, and *as*. Also playable on iPhone, iPad, or iPod, this Apple Watch game includes a real-time player-versus-player mode, or you can compete against others through the Game Center leaderboard. Snappy Word's dictionary is said to be made up of nearly 4,000 words.

Watch Quest! Heroes of Time

Even before Apple Watch debuted in the spring of 2015, buzz amassed in the gaming world over an adventure title called Watch Quest. Players first choose characters on the companion app for iPhone, equip them with gear, and then have them embark on a quest on Apple Watch — filled with engaging in combat, solving puzzles, and hunting treasure. Starring a male or female protagonist, this colorful game includes passive play — slower-paced and easier quests — and more difficult active play, including battling monsters and foraging for items. Now called

Watch Quest! Heroes of Time, it's free to download and comes with one training campaign, but you have to pay for additional quests and characters from within the iPhone app.

Spy_Watch

One of the more intriguing offerings sneaking onto Apple Watch is from UK-based indie developer Bossa Studios — best known for Surgeon Simulator and I Am Bread. In Spy_Watch, you're the head of a spy agency that's seen better days. To turn the agency around, you must train spies, send them on secret missions, and earn money to improve their abilities.

According to the developer, this ambitious Apple Watch game centers around short bursts of real-time alerts between you and your agent in the field, which take place in two- to five-second rounds. As the agency head, you need to make strategic decisions on the fly that will ultimately affect the success of the missions — be they related to time, resources, or locations. No guarantees you'll look as suave as James Bond while tapping on your wrist.

Rules!

With the iOS version taking home multiple award nods and an official App Store pick, Rules! was one of the first games announced for Apple Watch. And it still remains popular today. In case you haven't played the hit action puzzler from The Coding Monkeys, you're tasked with tapping tiles based on the rules for the level, such as "Tap numbers in descending order," "Tap green" (tiles), or "Tap animals" (as opposed, to, say, monsters or robots). When you advance to a new level, the rules revert to an older one, and you must remember what they were. Along with its 100+ memory-testing levels, there's also a daily "brain workout" mini-game challenge, an animated sticker pack, multiple difficulty settings, and other noteworthy features.

Cosmos Rings

From Square Enix, the same company that introduced Final Fantasy and Secret of Mana, comes a mobile adventure with charming, pixelated graphics. This simple-to-pick-up yet surprisingly deep role-playing game (RPG) offers a unique combat mechanic designed for the small screen: While basic attacks are automatic, you must enable powerful skill relics, strengthen your sword, and initiate a chain sequence to battle your way through this dark world. Be aware: It costs $8.99 but is well worth it. Need to go back in time to retake a mission? Simply twist the Apple Watch's Digital Crown to turn back the clock!

5

The Part of Tens

IN THIS CHAPTER

» Covering ten must-try features of
Apple Watch

» Diving deeper with additional tips
and tricks

» Impressing your friends with your
new wrist-mounted gadget

Chapter **13**

Ten Cool Things to Do with Your Apple Watch

Apple Watch has hundreds of use-case scenarios — many of which are provided by Apple and its built-in features — while third-party apps extend the functionality of this smartwatch even further (as evidenced in Chapters 11 and 12). But if you're like most people, you won't have time to go over *everything* Apple Watch has to offer. As I mention earlier in this book, it's estimated we only use about 10 percent of what our gadgets can do — until someone shows you what you're missing.

So, in this final chapter of *Apple Watch For Dummies,* I isolate ten must-try features of Apple Watch. Consider it a purely subjective list of my favorite things about the device.

Activity

Many smartwatches and fitness bands can report on your performance while working out, but Apple Watch is always calculating what you're doing — or not doing. The innovative Activity app and its three rings — for Move, Exercise, and Stand — does a stellar job of giving you an idea of your overall physical activity.

Press the Digital Crown button or lift your wrist and say "Hey Siri, Activity" and then take a gander at your progress:

>> The reddish-pink Move ring shows how many calories you've burned by moving around during the day.

>> The lime-green Exercise ring is for minutes of brisk or intense activity you've completed that day.

>> The baby-blue Stand ring gives you a visual indication of how often you've stood up after sitting or reclining.

Your goal is to complete each ring each day. The more solid each ring is, the better you're doing. Plus, you can swipe around inside the app for a numerical look at your performance.

You can also change your goals per day in case they're too ambitious for your lifestyle, or you can bump them up for an added challenge. The companion Activity app for iPhone shows you additional information, including a historical look at your Activity levels. Plus, every Monday, you should receive a summary report on your Apple Watch about your activity and goals. And there's the new Trends feature, too, for an even deeper look at your activity history (synced with iPhone). Chapter 8 offers a closer look at the Activity and Workout apps.

Apple Pay

Using your watch to buy things at retail is incredibly convenient. Even if you don't have your iPhone around, you can wave your wrist over one of those contactless terminals at the checkout counter or at an Apple Pay–compatible vending machine and the transaction is completed. And securely.

To buy something using Apple Pay on your Apple Watch, follow these steps:

1. **Double-tap the side button on Apple Watch, which opens Apple Pay.**

    Apple Pay uses your default card in the Wallet app, but you can change it to something else if you like. See Chapter 10 for more on the Wallet app.

2. **Hold the watch up to the contactless terminal, and you should hear a tone and feel a slight vibration — both of which confirm the payment has been made.**

    That's all there is to it. Apple Pay uses near field communication (NFC) technology inside of Apple Watch to make the *digital handshake* with the retailer's contactless terminal.

Apple Pay is supported by many banks and financial institutions as well as many thousands of retailers. But remember, you need to set up Apple Pay first on your iPhone if you haven't done so already. See Chapter 10 for more on Apple Pay.

Hotel Key

Free apps such as Marriott Bonvoy let you tap your watch on your hotel door to gain entrance. No more fumbling for the key card or having it demagnetized because you had it in your pocket with your smartphone.

If you've got the free app installed, tell someone at the check-in desk at a Marriott, Sheraton, Westin, W Hotel, Meridien, St. Regis, Element, or Aloft. Keep in mind that support for Apple Watch likely won't be available at all of these hotels and resorts or rolled out at the same time.

In the near future, expect many similar apps to let you into your car — instead of needing a large key fob — or to enter public transit stations, including bus depots and train terminals. Perhaps soon, Apple Watch will let you walk through your front door at home (with Wi-Fi smart deadbolts) or into your office by tapping your wrist on a card reader. See Chapter 11 for other third-party Apple Watch apps to check out.

Walkie-Talkie

In case you haven't given it a shot yet, Walkie-Talkie is a fun way for two Apple Watch wearers to communicate between themselves. As you might expect, Walkie-Talkie lets you quickly chat with someone, wrist to wrist, using your voice.

To get going, you and the person you'd like to talk with need to set up the Face-Time app on your iPhone. This app enables you to make and receive FaceTime audio calls. See Chapter 5 for info on that.

To use the Walkie-Talkie app on your Apple Watch, follow these steps:

1. **Open the Walkie-Talkie app (it's yellow and black) on your Apple Watch.**

2. **Press the yellow + sign and choose a contact.**

   Wait for your friend to accept the invitation. The contact card remains gray and reads "invited" until your friend accepts.

3. **After your friend accepts, his or her contact card turns yellow.**

   You and your friend can now talk instantly.

4. **Touch and hold the talk button, and then say something.**

   Now your friend can hear your voice and talk with you instantly.

5. **To talk over Walkie-Talkie, touch and hold the talk button, then say something; when you're done, let go.**

   Your friend instantly hears what you said. To change the volume, turn the Digital Crown.

REMEMBER

Apple Watch has Wi-Fi or cellular support on some models, but that doesn't mean you can surf the web; Apple Watch doesn't come with a web browser; it uses Wi-Fi only to move or sync data between it and your iPhone. That's probably not a bad thing, given the fact Wi-Fi eats up valuable battery life pretty quickly. Also remember that Walkie-Talkie requires that both people be running watchOS 5 or newer.

Music Playback

Many people who exercise rely on music to help keep them entertained and motivated. You might not want to bring a large iPhone with you on a jog or run, so Apple thankfully lets you sync some music to Apple Watch — up to 2 gigabytes, or about 500 songs.

(Otherwise, you'll need an Apple Watch with LTE cellular support, pay for a monthly data plan, and a subscription to the Apple Music service!)

To sync music to your Apple Watch, follow these steps:

1. **Connect your Apple Watch to your PC or Mac via its USB charger.**

   Use the magnetic charger that shipped with your Apple Watch.

2. **On your iPhone, open the Apple Watch app.**

3. **Under My Watch, scroll down and tap Music, followed by + Add Music.**

   Music is downloaded when Apple Watch is connected to power and placed near your phone.

4. **Tap to select something to listen to on Apple Watch.**

   Unplug the Apple Watch from the computer when the sync is complete.

After you have songs stored on your Apple Watch, follow these steps:

1. **Open the Music app and select from one of a few options: Now Playing, Listen Now, Radio, Library, Search on iPhone.**

   When you're inside these subsections, you will see options for Artists, Albums, Songs, Playlists, and Downloaded.

2. **Tap one of the options to begin streamlining music.**

   You're prompted to pair a Bluetooth-enabled headset or headphones to hear the music. The Apple Watch screen shows you what's playing on your watch or iPhone.

Apple Watch also acts as a remote control for an Apple TV connected to a TV or for playing music on an iPhone or iPad. See Chapter 9 for more on these and other media- and entertainment-related tasks.

Maps

Because Apple Watch is always on your wrist, it's a conveniently placed screen for getting directions. Apple Watch can give you turn-by-turn directions by tapping into your nearby iPhone's GPS chip, and you should see the overhead map on your watch, including a blue dot for your location, a red pushpin for the destination, and the path to take to get there quickly. Apple Watch gently vibrates to tell you when it's time to turn left or right. Some Apple Watch models have an integrated compass, too, for even more accurate directions.

To use the Maps app on your Apple Watch, follow these steps:

1. **Press the Digital Crown button to go to the Home screen.**

2. **Tap the Maps app.**

   You can also raise your wrist and say "Hey Siri, Maps." Either action opens the Maps app. You will see options to Search for something, bring up your Location, and scroll down to access Recents (places you've recently looked up or traveled to).

3. **To Search and say an address — such as a nearby shopping mall — or on Apple Watch Series 7, you can type in words with the onscreen keyboard.**

   If you make a mistake, tap Clear. If you're happy with what you requested, continue to the next step.

4. **Tap Start to begin the turn-by-turn directions.**

   You now see and feel when it's time to turn left or right when nearing an intersection — whether you're on foot or in a vehicle. Your iPhone also shows you information if you want to peek at a bigger screen (safely) or hand it to a passenger. See Chapter 6 for more on the Maps app.

Digital Touch

Many smartwatches on the market offer similar features, such as seeing who's calling or texting, calculating fitness information, or getting directions to a destination. But Apple Watch offers a few unique watch-to-watch communication options — collectively referred to as Digital Touch (see Chapters 3 and 5 for more on this feature). Here are three examples of them:

>> **Sketch:** Draw something with your finger, and the person you're sending it to sees it animate on his or her Apple Watch.

>> **Tap:** Send gentle (and even customizable) taps to someone to let that person know you're thinking about them.

>> **Heartbeat:** Want to tell a special someone you're thinking about them? A romantic way to do it is to send your heartbeat, but they'll need an Apple Watch to feel it! Your built-in heart rate monitor is captured and sent to someone special.

To send a heartbeat with your Apple Watch, follow these steps:

1. **Open the Messages app and tap someone.**

   Start a new message or reply to an existing conversation.

2. **Tap the blue-and-white icon that looks like two fingers on a heart.**

   You see a black screen, ready for your fingers. And you can tap the top-right corner to change colors.

3. **Press and hold your fingers on the screen, and you'll feel it pulse.**

   When you feel the pulsations stop, you can lift your fingers up, and your heartbeat is sent to the recipient.

Siri

Because Apple Watch was designed for quick interactions and to get information wherever and whenever you need it most, the best way to interact with your watch is by your voice. Providing you're in a place where you can talk freely, speaking into your watch's microphone is a fast, accurate, and convenient method for getting what you want when you want it.

If you recall, you can use Siri in three ways on Apple Watch:

» **Digital Crown:** Press the Digital Crown button and wait to see the little bars jumping up and down near the bottom of your screen. This confirms Siri is "listening" to you.

» **Voice Activation:** Simply say "Hey Siri" and your personal assistant on your wrist will be ready for your question or command. If there is another Apple device nearby, like an iPhone or iPad, it may defer to that device first. Enable this feature in the Watch app on your iPhone (Siri ⇨ Listen for Hey Siri).

» **Raise to Speak:** Raise your wrist and say "Hey Siri," followed by your command or question. Or you can go into and change the Settings of Apple Watch to enable simply by raising your wrist to activate your personal assistant.

WATCHES GALORE!

Most companies that release a smartwatch have one or two models, but Apple Watch is available in multiple sizes, several materials (aluminum, stainless steel, and 18-karat gold), in multiple case colors, and with various band colors, materials, and styles to choose from. Clearly, Apple has thought this through!

Even with all the options, the user experience will be similar between all the versions because features, interfaces, and apps are the same for all of them. The following is a quick summary of the six current options (at the time of writing this second edition of the book):

Apple Watch Series 7: Launched in late 2021, Apple Watch Series 7 is the largest of all Apple Watches, with a screen about 20 percent larger, but because there are thinner bezels (borders), it's not much bigger in overall size compared to its predecessors. Apple also added a tougher screen and faster charging.

(continued)

(continued)

Apple Watch Series 8: The latest in the Apple Watch family is Apple Watch Series 8, which debuted in 2022, with several new features, including temperature sensing (for deeper insight into women's health), Crash Detection, and more.

- **Apple Watch Ultra:** The biggest and strongest Apple Watch is designed for outdoorsy types and extreme athletes. Along with added durability, it houses a super long battery, customizable Action button, advanced GPS location sensors, and Crash Detection, to name a few features.

- **Apple Watch SE:** Much like the less expensive iPhone SE, Apple Watch SE is meant to give you premium features at a more affordable price. It includes a great-looking Retina display, advanced sensors to track your movement and sleep, and more. It was updated in 2022 (second-generation Apple Watch SE).

- **Apple Watch Nike+:** Ideal for fitness types who like the Nike brand, this special edition Apple Watch Series 4 (and special loop band) was designed to be your running partner and synchronizes with the Nike Run Club app and Nike Training Club app.

- **Apple Watch Hermès:** A partnership between Apple and Hermès, this fashion-centric watch includes bold, colorful (and extra-long wraparound) leather bands and an exclusive watch face.

See www.apple.com/watch for more information on the Apple Watch collections and some accessories.

For both of these options, you should get what you need within a second or two, but remember, you need your iPhone nearby because your request is quickly sent to Apple's servers to process it. See Chapter 7 for more on using Siri to help you complete tasks with your Apple Watch.

Gaming

It's a huge understatement to say Apple Watch is an unproven video game platform. But given Apple's track record with iOS devices — not to mention a passionate app development community eager to take advantage of this new real estate on the wrist — gaming might be the secret "killer app" of Apple Watch.

You're in line at the supermarket, and you want to kill some time by dunking a few virtual baskets by tapping your watch screen. Or you're on the train to work, and you want to use your fingertip to slide letter tiles on a board to create a word. Or perhaps you're walking down the street, and you feel a tap on your wrist — an alert that someone is invading your village, and you've got to decide what to do.

Just as the smartphone and tablet have become viable gaming platforms in a very short period of time — even pumping out such iconic games as Angry Birds and Flappy Bird — Apple Watch can introduce fresh gaming experiences on a device we always have strapped to our wrists.

Chapter 12 offers a look at ten games available for Apple Watch, but the App Store — accessible on the companion Apple Watch app on iPhone — has thousands more to choose from.

Mute Alerts with Your Palm

Now this is a handy feature! Even if you've enabled sounds to emanate out of your Apple Watch, there are times where you'll want to keep it from beeping when in public. Whether you're at the library, on a date, or riding with others in a jammed elevator, if your wrist-mounted device chimes to remind you of something, simply cover the display with your hand for three seconds or so, to instantly mute any sounds.

To enable this feature, visit the Apple Watch app on your iPhone and then go to My Watch ⇨ Sounds & Haptics ⇨ Cover to Mute. This option should be at the bottom. Make sure it's enabled (green tab), and you're good to go.

Index

X

Y

Z

About the Author

Marc Saltzman is a prolific journalist, author, and TV/radio personality who specializes in consumer electronics, business technology, interactive entertainment, and Internet trends. Marc has authored 16 books since 1996 and currently contributes to more than 20 high-profile publications, including *USA TODAY*/Gannett, AARP, Common Sense Media, *Costco Connection*, *Toronto Star*, Zoomer, Homefront (HF), and others. Marc is the host of the *Tech Impact* television show, seen on Bloomberg TV and Fox Business. Marc also hosts "Tech It Out," a nationally syndicated radio show, heard on more than 100 U.S. radio stations and is a podcast, too. *Apple Watch For Dummies* is Marc's second *Dummies* book. He also penned *Siri For Dummies*. Follow Marc on Twitter: @marc_saltzman or Instagram (@marcsaltzman). Or visit his website at www.marcsaltzman.com.

Dedication

This book is dedicated to my extraordinary parents, Stan and Honey Saltzman. A heartfelt thank you for your never-ending support, guidance, and love.

Author's Acknowledgments

I'd like to acknowledge all the talented folks at John Wiley & Sons because this book wouldn't have happened without their professionalism, patience, and knowledge (they're hardly Dummies!). In particular, I'd like to thank Steve Hayes, Elizabeth Stilwell, and Christopher Morris.

Publisher's Acknowledgments

Acquisitions Editor: Elizabeth Stilwell

Project Editor: Christopher Morris

Copy Editor: Christopher Morris

Technical Editor: Thomas Egan

Production Editor: Tamilmani Varadharaj

Cover Image: © monkeybusinessimages/ Getty Images